Contents

Illustrations

General maps

Text figures

Roman urban defences in the west

Roman urban defences in the west

edited by John Maloney
and Brian Hobley

A review of current research on urban defences in the Roman Empire with special reference to the northern provinces, based on papers presented to the conference on Roman Urban Defences held at the Museum of London on 21–23 March 1980

1983

Research Report 51

The Council for British Archaeology

ISBN 0 906780 23 3

Designed by Allan Cooper FSIA and Henry Cleere

Published 1983 by the Council for British Archaeology,
112 Kennington Road, London SE11 6RE

Printed by
Henry Ling Ltd, The Dorset Press, Dorchester

British Library Cataloguing in Publication Data

Roman urban defences in the West.—(CBA research
report ISSN 0589-9036; 51)
1. Fortification, Roman—Europe—Congresses
I. Maloney, John II. Hobley, Brian
III. Council for British Archaeology
IV. Series
623´. 12´0936 DE61.M5

ISBN 0-906780-23-3

Contributors

D Baatz, Saalburgmuseum, Saalburg-Kastell, 638 Bad Homburg vd Hl, Federal Republic of Germany

T F C Blagg, University of Canterbury, Rutherford College, Canterbury, Kent CT2 7NX

R M Butler, Royal Commission on Historical Monuments (England), The White House, 9 St Marks Grove, Rawcliff, York YO3 6TS

J Casey, University of Durham, Department of Archaeology, The University, Co Durham DH1 3NU

C Daniels, University of Newcastle upon Tyne, Department of Archaeology, Newcastle upon Tyne NE1 7RU

M Fulford, University of Reading, Department of Archaeology, Whiteknights, Reading RG6 2AA

M Hassall, University of London, Institute of Archaeology, 31–34 Gordon Square, London WC1H 0PY

H Hellenkemper, Römische-Germanisches Museum, Roncalliplatz 4, Köln, Federal Republic of Germany

B Hobley, Museum of London, Department of Urban Archaeology, London Wall, London EC2Y 5HN

S Johnson, Inspectorate of Ancient Monuments and Historic Buildings, Fortress House, 23 Savile Row, London W1X 2AA

M J Jones, Lincoln Archaeological Trust, The Sessions House, Lindum Road, Lincoln LN2 1PB

J Maloney, Museum of London, Department of Urban Archaeology, London Wall, London EC2Y 5HN

J Mertens, Université Catholique de Louvain, Av des Genets 32, B 1970 Wezembeek, Belgium

G Precht, Rheinisches Landesmuseum, Colmanstrasse 14–16, 5300 Bonn, Federal Republic of Germany

M Todd, University of Exeter, Department of History, Queen's Building, The Queen's Drive, Exeter EX4 4QH

J Wacher, University of Leicester, Department of Archaeology, The University, Leicester LE1 7RH

G Webster, The Old School House, Chesterton, Harbury, nr Leamington Spa, Warwickshire CV33 9LF

Introduction

Urban defences of the Roman period are amongst the most impressive monuments to have survived into modern times and, in some instances, they determined the extent of towns until well into the Middle Ages. Hence these circuits have been subject to study from an early date, but the standard of investigation and presentation of the results has varied considerably. Further, although there are many publications on Roman military defences (eg Limes Conference reports), assessments of urban defences have been of a rather piecemeal character. Since in recent years archaeological excavations have provided much more evidence, many researchers felt that a review devoted specifically to the topic—for the first time—was long overdue. To this end a conference on 'Roman urban defences in the north-west provinces' was held in London on 21–23 March 1980, jointly organized by the Museum of London and the Council for British Archaeology. Sixteen papers were given by speakers from Britain, Belgium, and Germany, to an audience of over 120 people. Subsequently, Charles Daniels was prevailed upon to submit a paper on Roman Africa, and the title of the publication was changed to 'in the west' which now reflected its scope.

The seventeen papers published in this volume are divided into four sections: the Mediterranean and northern provinces, and Rome; Britain; military architecture and weapons; conclusion. A series of maps (Figs I–V) placed at the head of each relevant section feature all the place-names mentioned in the papers.

John Maloney
Brian Hobley

Acknowledgements

Our thanks must go first to the speakers from Germany, Belgium, and Britain, who submitted their papers for publication with commendable promptness following the conference. We also wish to acknowledge the guidance of Professor John Wilkes and John Wacher in planning the conference programme. Colleagues in the Museum of London and the Council for British Archaeology assisted with the arrangements and special mention must be made of Jean Coleman, Wendy Rix, and Henry Cleere. The Director of the Museum of London, Max Hebditch, and Vanessa Jones, Crispin Jones, Diana Twells, and Wendy Fakes of the Department of Urban Archaeology organized receptions which were much appreciated. The exhibition mounted for the conference owed much to the work of Trevor Hurst, Jon Bailey, Jan Scrivener, and Chris Green, and was generously sponsored by European Ferries.

We are indebted for assistance with the preparation of this volume to the following members of the Department of Urban Archaeology: typists Diana Twells and Wendy Fakes, Richard Lea for compiling the maps which involved the recording and processing of information by computer, and for work on the illustrations, and Clare Thorne for the cover drawing. Friedrike Hammer and Stephen Johnson kindly agreed to translate papers, from German and French respectively. Finally, we are grateful to Valerie Horsler and Henry Cleere of the CBA for guiding the volume through to publication.

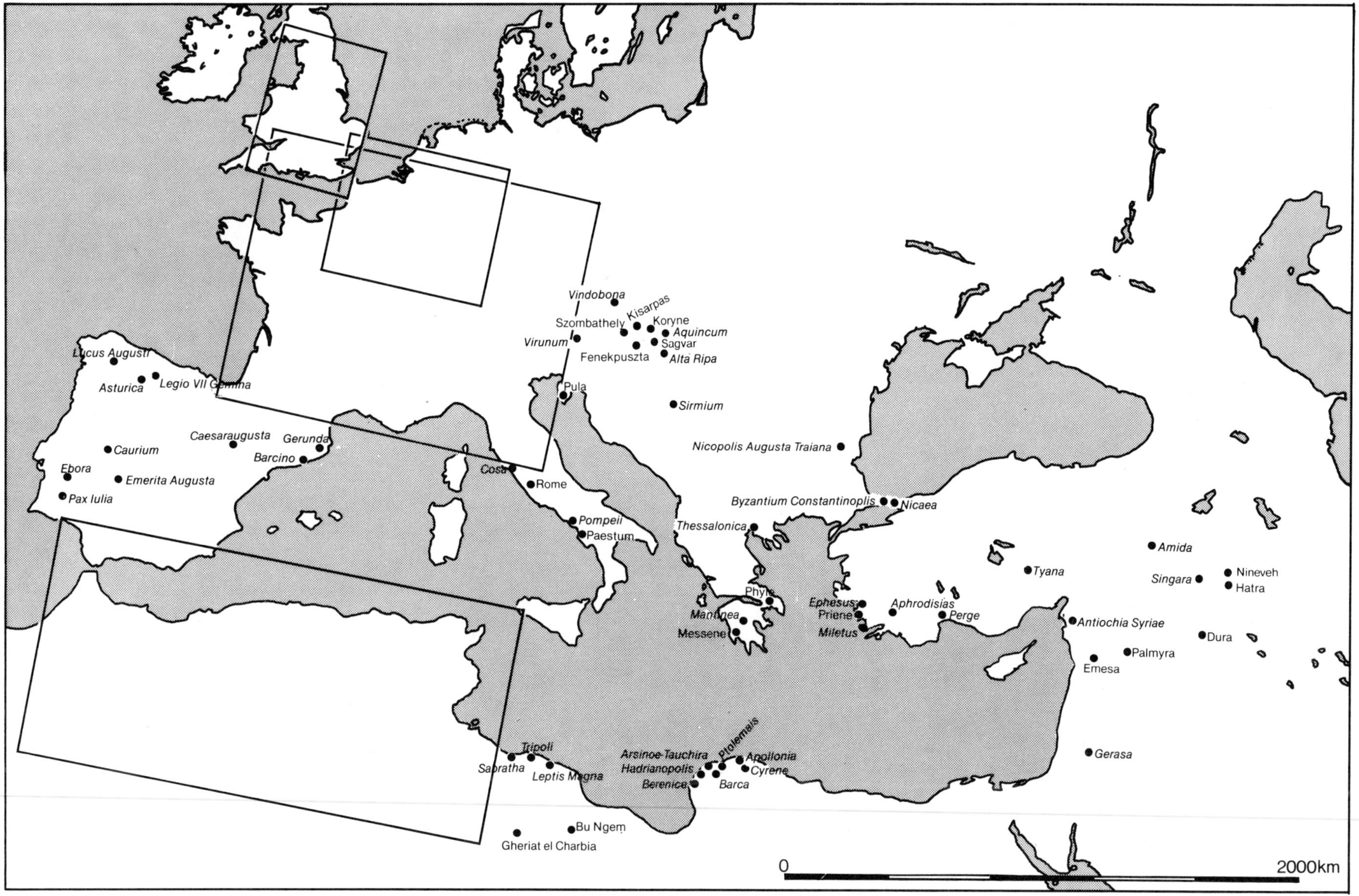

Fig I The Roman Empire

PART I MEDITERRANEAN AND NORTHERN PROVINCES, AND ROME

The origins and character of Roman urban defences in the west

Mark Hassall

The earliest urban defences in the north-west provinces of the Roman empire were those of the *coloniae* of southern France and Spain which had as their prototypes the fortifications of the Republican and Augustan *coloniae* of northern Italy. These in their turn followed the style of defensive architecture established in Magna Graecia and the Hellenized world of the Aegean and eastern Mediterranean lands. Typical of such Hellenistic fortified enceintes is Perge (Marsden 1969, 155, diagram 2, 163, diagram 14; Lawrence 1979, see index) with its projecting rectangular artillery towers in front of a horseshoe shaped gate court (Akurgal 1970, 332, fig 163).[1]

The exact date of the Hellenistic defences at Perge is not known but they will have been roughly contemporary with those of the *coloniae* at Cosa, 140 km north of Rome, founded in 273 BC (Brown 1980, figs 16–17 and town plan), with its eighteen square projecting towers built of massive polygonal masonry. Closer in feeling are the towers at Pompeii either built to defend the colony founded on the site by Sulla in 80 BC or belonging to the pre-existing Samnite city (eg Marsden 1969, 161, diagram 11, but note the top of this tower has been restored). And half a century later come the Augustan *coloniae* of north Italy such as Turin (founded *c* 25 BC) with the imposing Porta Palatina (Frigerio 1934/5, 133–42; Schulze 1909, 339–41) or Aosta (founded at the same time) whose circuit wall can still be traced (Richmond 1969) though the remains lack parapet and crenellations. As the type site of a late Republican *colonia,* it is worth quoting Richmond's description of the defences of Aosta (reproduced by permission of the Visitors of the Ashmolean Museum):

> The town-wall is just over 6 feet thick, reduced by three offsets to some 5 feet at the top. It stands 21 feet high to rampart-walk. It is faced with squared masonry set in regular courses, with a core of mortared cobbles and rubble. The rampart-walk is marked by a string course . . . The wall had no earthen rampart behind it. Reinforcement against battering rams was provided by buttresses 6 to 4 feet wide and 10 feet deep occurring at every 40 feet, a rare feature matched at Side and within other Pamphylian towns. The towers are rectangular, 30 feet across the front and 33 Roman feet from front to back: they project from the wall at front and back but rather more towards the front, where the ground floor wall is again thickened against battering rams to match the town walls. They are three storeys high with three windows on each face on the second floor, and the same, except that a door takes the place of a window, at the first floor. The ground floor was hollow and each floor was board. The top was presumably crenellated in the manner of conventional towers on contemporary mosaics or water heaters.

Aosta was provided with four gates (also described by Richmond). The north and south gates consisted of single portals only but the more important east and west gates were provided with passageways for pedestrians flanking their single arched carriageways and had gate-courts built behind them (Frigerio 1934–5, 151–3; Schulze 1909, 293–7).

Built according to the same Hellenistic traditions were the walls of the late republican or early imperial *coloniae* of southern France (see below) and Spain such as Emerita and Saragossa (Richmond 1930b, 100–3; 1931, 94–5; Schulze 1909, 304, fig 6). In some cases indeed the provincial defences might be closer to Hellenistic than north Italian prototypes; thus Wheeler (1962, 182) long ago compared the gates of Arles and Fréjus and Aix-en-Provence set at the rear of half-moon shaped gate courts flanked by circular towers, with one of the gates at Mantinea following a suggestion already put forward at the end of the last century. (There is also a certain resemblance to the Hellenistic gate at Perge already mentioned, only there the gate court is narrower and more enclosed at the front.) Like the premier *colonia* in Provence, Narbo (founded 118 BC), these towns were bastions of empire set amongst hostile tribes (cf Cicero, *Pro Fonteio*, 5.13) and it is hardly surprising that they were furnished with defences. The *coloniae* north of the Provincia like Lyons in Gallia Comata or Xanten or Cologne (see the papers by Precht and Hellenkemper, this volume) in the Rhineland were also defended by walls early in their history and for the same strategic reasons. Two of these sites, however, both in Switzerland, have defences whose design departs in a curious way from the standard Hellenistic pattern. At Augst (a Caesarian colony, though the date of the walls is unknown) and Avenches (raised to colonial status under Vespasian) the walls are furnished with semicircular towers built against their inside face instead of projecting as normally. Artillery mounted on such towers could not have provided flanking fire and it is difficult to see why they were built in this way.

In contrast to the semi-military *coloniae* the towns built by the tribal *civitates* were not walled. What was the need for walls when even the city of Rome could dispense with them, being protected indeed, like the provincial towns themselves, by the very frontier defences of the empire (cf Aelius Aristeides, *To Rome*, section 79)? Apart from some minor defended settlements in the Rhineland (Butler 1959) the only real exceptions may have been the chief towns of *civitates foederatae,* where the possession of defences could be seen as, literally, a status symbol. Autun (Augustodunum), the chief town of the allied Aedui, had walls whose gates still survive (Frigerio 1934/5, 122–8; Schulze 1909, 304–11). Rheims, Durocortorum, the chief town of the allied Remi, may also have had defences—of earth—the position of whose entrances were marked by triumphal archways as at the *municipium* of Verulamium. If so the town soon outgrew this early circuit, though in the late empire it contracted once more and the arched gates were incorporated into the late Roman defences.

In Britain, most surprisingly, the *colonia* at Colchester founded by Claudius was not at first defended, an omission for which the colonists were to pay dearly at the time of the Boudican revolt. Verulamium, the sole *municipium* in the province, as noted above, had original earthen defences. Winchester and Silchester (see Fulford, this volume) also had early earthen defences, convincingly explained by John Wacher who suggests that they may have lain in the territory of those tribal states that were entrusted to the supervision of the client king Cogidubnus (Wacher 1975, 256). The earthen defences of these four towns will probably have resembled those of contemporary forts and fortresses (Jones 1975) and this was the pattern adopted by the majority of towns in the province when they were defended in the 2nd century, a century or so before their counterparts across the channel (Wacher 1962). At Cirencester the earthen defences were provided with stone towers (Wacher 1975, 202). Elsewhere, if interval towers existed, they may have been of wood, in which case it will be reasonable to assume that they were incorporated into the body of the earth rampart as with contemporary military fortifications since a projecting wooden tower would be especially vulnerable to enemy attack. (Even stone towers will normally have been set in the body of the rampart, as at Cirencester, since a stone tower built in front of the rampart at the bottom would be isolated from it at the top by the sloping batter of the rampart's front. It is thus surprising to hear that at Lincoln there were apparently projecting timber towers (M J Jones, this volume).)

In the course of the 2nd century forts in Britain regularly received stone defences. Sometimes as at Trajanic Gelligaer in South Wales (Nash-Williams 1969, 90) the fort's stone wall was backed by an earth rampart retained at the rear by a second, presumably lower, wall, a translation into stone of the timber 'box' method of construction which had sometimes, especially in the Rhineland, retained the earthen defences of 1st century forts and fortresses. Sometimes, as, apparently, at Hadrianic Dover (Philp 1981, 20–3, fig 5) there may have been only a stone wall with no earth rampart behind it. Normally, however, stone interval towers were still built with their fronts flush with the stone wall although there was now no reason why they should not have projected like the standard Hellenistic and late Republican/Augustan artillery towers.

This digression into a discussion of military fortifications in Britain in the 2nd century has been given because they provided the model for urban defences in the province in the 3rd. During that century the earthen defences of the towns were refurbished by cutting back the front of the rampart in order to build a stone wall, but the opportunity was not taken to provide the defences with projecting stone towers at the same time. It was not until the 4th century that this anomaly was finally corrected and the defences of towns of Roman Britain were brought into line with the norm, established four centuries previously: the ditches immediately in front of the walls were filled in and a new wider ditch was dug at some distance from it, thus creating a strip of made ground on which projecting towers could be built (Corder 1955). In contrast, when eventually the towns of Gaul—or for that matter Rome itself—were given walls they naturally had projecting artillery towers from the start (Butler 1959; Richmond 1930a; Todd 1978).

By the 4th century, then, the majority of towns in the Roman west at last had defences and defences of a type that could be used in conjuction with torsion artillery. Yet in one respect all, or probably all, the cities of the Roman west fell behind the most sophisticated of the cities of the east in their defensive arrangements. This was in the provision of an outwork or *proteichisma* (cf Marsden 1969, 117–21; Lawrence 1979, 276–301), a low wall built on the inner scarp of the city ditch behind which artillery could be placed for use at short range. This was particularly important because the 'barrels' of Hellenistic and Roman catapults could probably not be depressed, or only slightly, so that the space immediately in front of the defences was in effect dead ground as far as artillerymen stationed on the main walls or towers were concerned. *Proteichismata* are known from a number of Hellenistic sites including St Blaise near Marseille (Lawrence 1979, 284–6) and the use of such outworks was revived during the 5th century (Crow forthcoming), most spectacularly at Constantinople, where the land defences were completed by 422 (Schneider 1937; Meyer-Plath & Schneider 1943). If Constantinople itself set the fashion then the absence of *proteichismata* is easy to explain since the revival will have come too late to affect the west. However, it seems likely that *proteichismata* were already being constructed during the 4th and even 3rd centuries, so that other reasons should be adduced to explain their presence in one area and not the other—such as the greater sophistication of Rome's eastern opponents in siege warfare. The earliest certain evidence for the revived use of a *proteichisma* probably comes from Singara in northern Iraq where the defences should date to before AD 360 (Oates 1968). An even earlier case may be Nicaea (Schneider 1938; Schneider & Karnapp 1938). Here the well preserved *proteichisma* is now thought to date from the 13th century but may have had a predecessor contemporary with the city walls which are dated to the second half of the 3rd century (*CIG* 3748). At all events the original D-shaped bastions are provided with posterns whose presence could best be explained by the need to man an outwork.

If *proteichismata* were already in use in the 3rd and 4th centuries the question arises whether, after all, they really are absent from the western empire during the same period. The only certain evidence is a passage in Ammianus' description of the siege of Aquileia in 362 which refers to a crude outwork, 'a rampart before the walls covered with turf protected from all dangers those who lay in wait' (Ammianus, xxi.12.13—I owe this reference to Dr J G Crow). Finally it may be just worth drawing attention to the late Roman military posts at Junkerath, Bitburg, and Neumagen, all in eastern Gaul. These three are all fortified road stations built astride main roads which enter through one of the two gates and leave at the other (for plans see Anthes 1917, 103, figs 6–8). But, in addition to the main gates, in each case several of the towers appear to have posterns cut either through them or through the adjacent curtain wall which might just suggest the presence of some sort of outwork, as we have argued may have been the case with Nicaea. If artillery were used to defend these posts it could have come from nearby Boppard (Baudobriga) which housed the only unit of *ballistarii* stationed in the west (*Not Dig* (*Occ*) xli, 23).

Note

[1] For sites for which no specific reference is given in the text, see in general the articles in Stillwell *et al* 1976.

References

Akurgal, E, 1970 *Ancient civilizations and ruins of Turkey*, 2 edn

Anthes, E, 1917 Spätrömische Kastelle und feste Städte im Rhein und Donaugebiet, *Bericht der Römisch-Germanischen Kommission*, **10**, 86–167

Brown, F E, 1980 *Cosa: the making of a Roman town*

Butler, R M, 1959 Late Roman town walls in Gaul, *Archaeol J*, **116**, 25–50

CIG *Corpus Inscriptionum Graecarum*

Corder, P, 1955 The reorganisation of the defences of Romano-British towns in the fourth century, *Archaeol J*, **112**, 20–42

Crow, J G, forthcoming The walls, in *Tocra* (ed D J Smith)

Frigerio, F, 1934/5 Antiche porte di città italiche e romane, *Riv Archeol di Como*, **108–10**, 1–284

Jones, M J, 1975 *Roman fort-defences to AD 117, with special reference to Britain*, Brit Archaeol Rep, **21**

Lawrence, A W, 1979 *Greek aims in fortification*

Marsden, E W, 1969 Artillery and fortifications, in *Greek and Roman artillery: historical development* (ed E W Marsden), 116–63

Meyer-Plath, B, & Schneider, A M, 1943 *Die Landmauer von Konstantinopel*, pt 2, *Aufnahme, Beschreibung und Geschichte*, Deutsche Forschung-gemeinschaft: Denkmäler antiker Architektur, **8**

Nash-Williams, V E, 1969 *The Roman frontier in Wales*, 2 edn, rev M G Jarrett
Not Dig (Occ) O Seeck (ed), *Notitia Dignitatum (in Partibus Occidentis)*, 1876
Oates, D, 1968 Singara and its fortifications, in *Studies in the ancient history of northern Iraq* (ed D Oates), 97–106
Philp, B, 1981 *The excavations of the Roman forts of the* classis Britannica *at Dover 1970–1977*
Richmond, I A, 1930a *The city walls of imperial Rome*
———, 1930b The first years of Emerita Augusta, *Archaeol J*, **87**, 98–116
———, 1931 Five town walls in Hispania Citerior, *J Roman Stud*, **21**, 86–100
———, 1969 Aosta, in *Roman archaeology and art: essays and studies by Sir Ian Richmond* (ed P Salway), 249–59
Schneider, A M, 1937 The city-walls of Istanbul, *Antiquity*, **11**, 461–8
———, 1938 The city-walls of Nicaea, *ibid*, **12**, 437–43
Schneider, A M, & Karnapp, W, 1938 *Die Stadtmauer von Iznik* (Nicaea)
Schulze, R, 1909 Römische Stadttore, *Bonner Jahrb*, **118**, 280–352
Stillwell, R, MacDonald, W L, & McAllister, M H, 1976 *The Princeton encyclopedia of classical sites*
Todd, M, 1978 *The walls of Rome*
Wacher, J, 1962 A survey of Romano-British town defences of the early and middle second century, *Archaeol J*, **119**, 103–13
———, 1975 *The towns of Roman Britain*
Wheeler, R E M, 1926 The Roman town-walls of Arles: and a note on other Roman town-walls in Gaul and Britain, *J Roman Stud*, **16**, 174–93

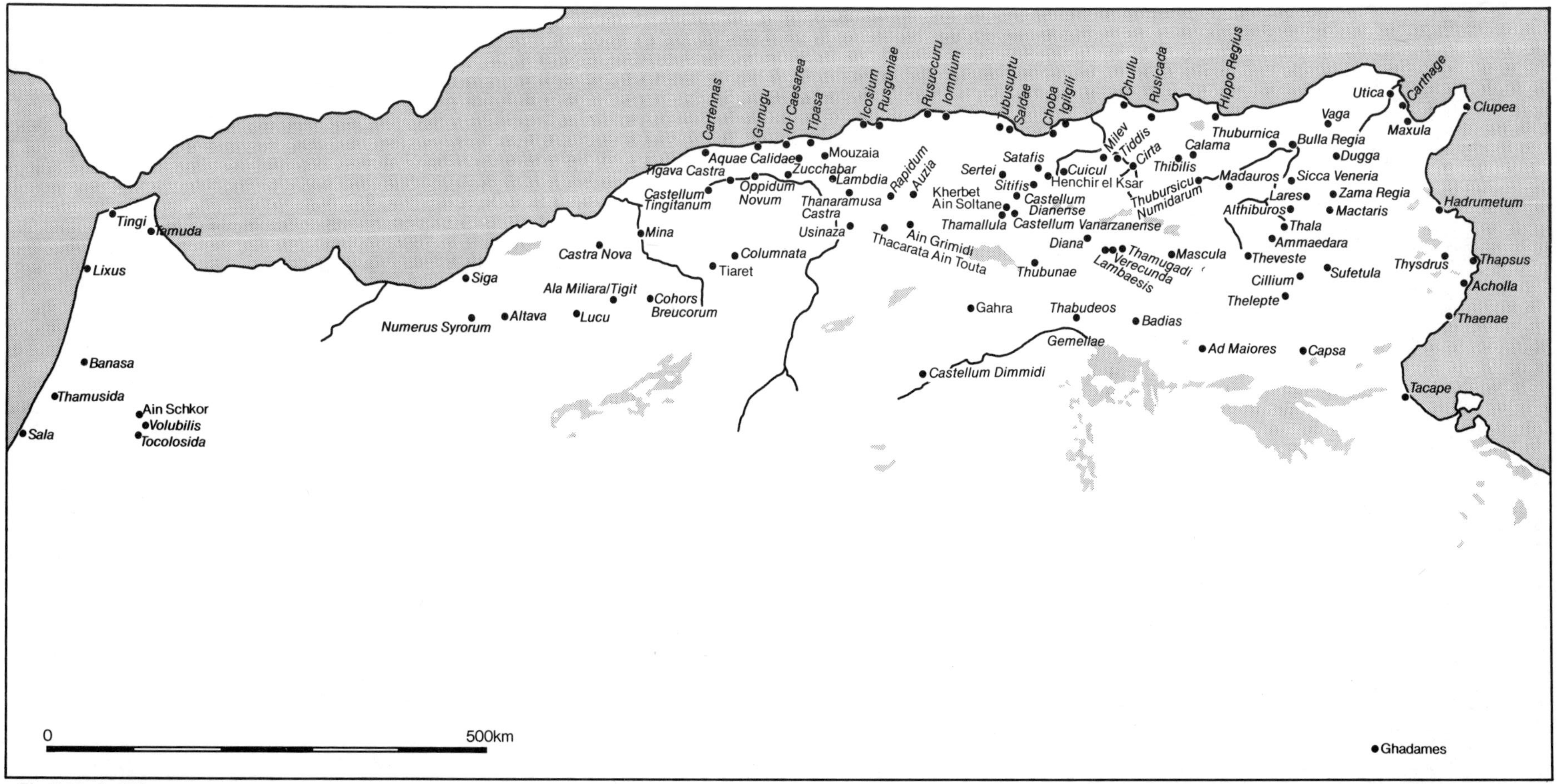

Fig II North Africa

Town defences in Roman Africa: a tentative historical survey

Charles Daniels

Introduction

The cities of Roman Africa were never a uniform group, which is not surprising when it is realized that Apollonia (Marsa Susa) is well over 1700 miles from Sala (Rabat), or rather further than Moscow is from London. Nor were their origins similar: the cities of the Pentapolis were Greek colonies, those west of the Syrtic Gulf, along the Mediterranean and Atlantic coasts, were mostly Phoenician and Carthaginian settlements, while inland cities such as Cirta, Dugga, and Volubilis were of native origin. In Roman times new towns were the result of veteran settlement or grew from fort *vici*, while in other cases older creations became *municipia* or *coloniae*, especially during the 1st and 2nd centuries AD.

The significance of the creation of a colony has recently been stressed, particularly by Février (1969) who argued, on the analogy of Augustan colonies in Gaul, that at the moment of its creation a colony received its walls. While not accepting that this was invariably the case (some older settlements already had defences while others do not appear ever to have received them), in what follows I have stressed the creation of colonies and *municipia* as of significance, particularly in Augustan times (when many colonies were planted in what was still very much un-Romanized Numidia and Mauretania) and through the 1st and early 2nd centuries (when archaeological evidence shows that at least some new creations were still receiving walls).

Detailed information, or sometimes even simple descriptions, are usually absent from the older records used by Gsell in his *Atlas*, but it is clear that projecting square and circular towers and recessed gates were employed from Augustan times, as they were at such sites as Fréjus, Arles and other cities in Narbonensis. The term often used to describe the construction of a wall is *en blocage*, which strictly speaking means rough stonework or wall-core. But it is not absolutely certain that Gsell's sources did not occasionally use it of roughly faced walling. I have sometimes used the term untranslated in what follows.

Although agreeing (somewhat reluctantly) to provide this survey, I cannot claim to have studied the defences of many African cities at first hand. Certainly I have visited many of the sites, but never with their defences as the prime object. This account, then, is something of an academic one and definitely light on actual description and comparison of the structures. It has been even further handicapped by the lack of easy access to several important publications, including the original descriptions of various sites. Amongst prime sources Stephane Gsell's *Atlas archéologique de l'Algérie* (Gsell 1911) is still the indispensable work. By contrast *The Princeton encyclopedia of classical sites* (Stillwell *et al* 1976), although undoubtedly of use, does not normally describe defences and, although comprehensive for Africa and Mauretania Tingitana, leaves very much to be desired for Caesariensis. Some unlikely sources, therefore, have been pressed into service.

Finally, I must thank Dr David Smith for discussing his unpublished work on the walls of Tocra-Arsinoë with me, Mr Eric Balley for his painstaking work in providing the illustrations, Mr John Dore for his comments on pottery, and Dr Stephen Johnson for discussing technical French architectural terms. Failure to comprehend remains my own.

The cities of Cyrenaica

Separated from both Egypt and Tripolitania by desert regions, the history of Cyrenaica was different from its neighbours. Here, alone in Africa, were cities of classical Greek origin which, as part of the Ptolemaic Empire, partook fully of the Hellenistic culture of the eastern Mediterranean before they passed to Rome. Not surprisingly they form a coherent group and are best treated as such.

On the death of Alexander the Great the Pentapolis fell to Ptolemy I, although for the next 70 years or so it enjoyed virtual independence. Ptolemy's stepson Magas, having first dismantled the city walls of Cyrene, later provided a new circuit (Goodchild 1963; White 1976a, 253). Apollonia, too, he surrounded with walls sometime in the first 25 years of the 3rd century BC (Goodchild 1963; White 1976a, 71; 1976b, 85–155). Lengths of this have been destroyed by the sea, but the landward section of *c* 1000 m survives, including portions of the western gate and nineteen projecting towers, two of which were circular, the rest rectangular. The whole was built of well cut and finely dressed monumental ashlar, sometimes with drafted margins, as can today best be seen on the acropolis hill at the south-eastern corner of the city. The one surviving gate is a modest passageway through the walls, protected by an imposing circular tower. However, it is unlikely that this was the principal entranceway to the city, so one or more gates must have been destroyed by the sea.

The next step in the development of city defences in the Pentapolis was taken by Ptolemy III, who married Magas' daughter Berenice. Euesperides was by now most probably suffering from the silting-up of its harbour (Goodchild 1952; Kraeling 1962, 431, fig 6; Brogan 1976, 320). In *c* 247 Ptolemy moved its population some 2 km and established a new city which he named after his consort. This city was defended by a strong circuit of walls, some of the foundations of which have recently been excavated (Fig 1). Like those of Apollonia and the acropolis at Cyrene they show the wall to have been constructed as a series of salients and re-entrants, although the masonry itself has not survived (Goodchild 1962; Lloyd nd, 43–52).

At the same time the old port of Barca was renamed Ptolemais and fortified with a truly Hellenistic circuit of walls some 5800 m (6300 yards or over 3½ miles) in length (Kraeling 1962, 51–62; Brogan 1976, 742), which not only enclosed the newly laid out city, but ran back inland up on to the lower spurs of the Jebel Akhdar (Fig 1). On the plain the construction was of first-rate quality, especially the Tauchira Gate (Fig 4). This was a single passageway defended by projecting rectangular towers, built of well cut monumental ashlar smoothly dressed with marginal drafting. The quality of this workmanship was superior to that displayed elsewhere in the Pentapolis cities. Square, projecting towers again reinforced the curtain, especially at the gates. On the escarpment spurs the wall has been less well studied, but it appears to have been of somewhat rougher quality.

A third coastal city, Tauchira (Brogan 1976, 886; also information D J Smith), was renamed Arsinoë, after Ptolemy's mother, and provided with defences of the same high quality (Fig 1). These enclosed an overall area measuring about 650 m in each direction. Lengths of the original work still survive on the landward sides, including

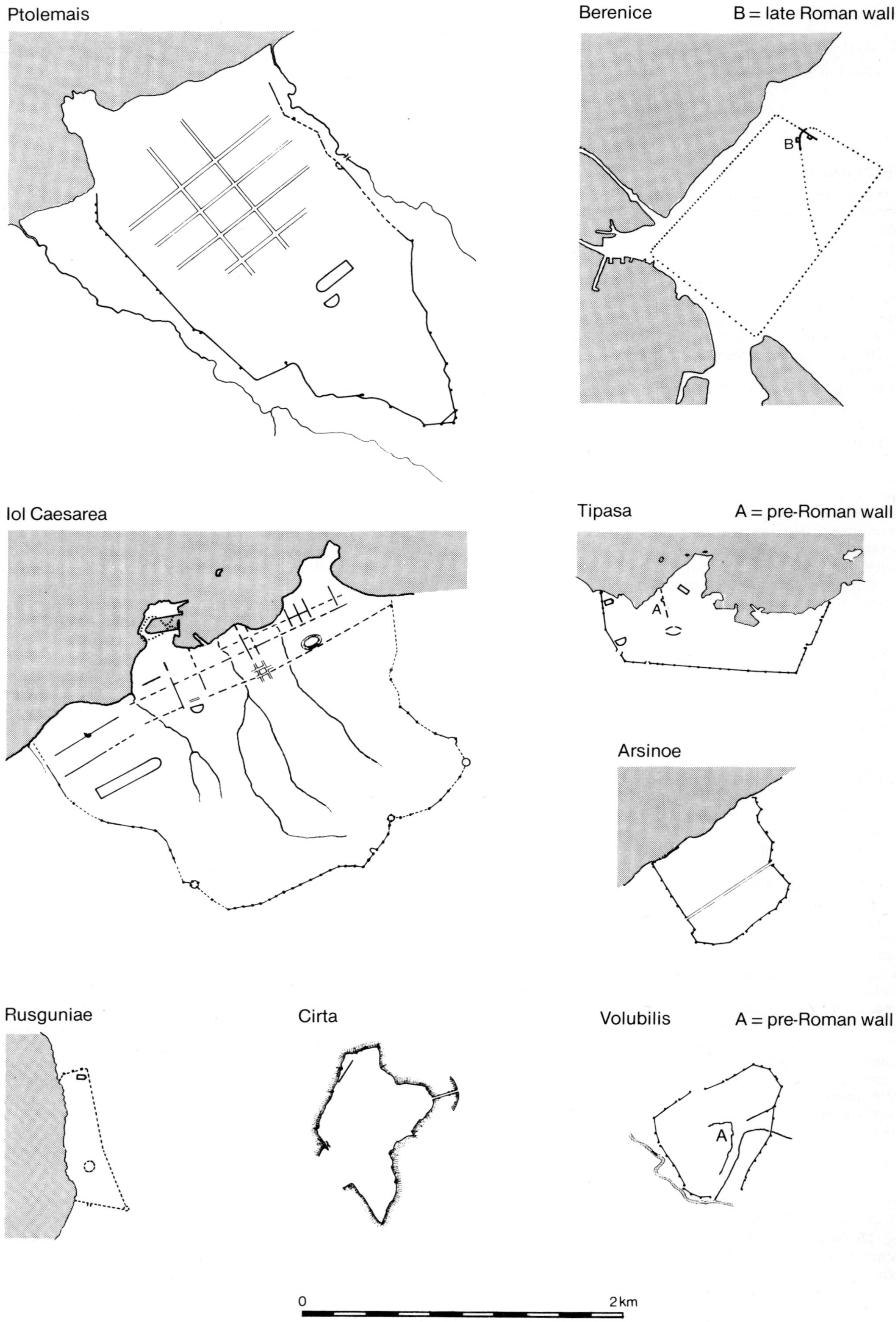

Fig 1 Roman towns in North Africa

the east gate, a single portal with unusual cut-water shaped fronts to its projecting towers.

These impressive city walls, built in the contemporary Hellenistic manner, remained in use throughout the next six or seven centuries; of sophisticated design and optimum construction they were not bettered, even in Byzantine times; enclosing copious areas they never needed to be demolished for later Hellenistic or Roman expansion. In fact, it was decreasing populations in later antiquity which led to an abandonment of the walls of Ptolemais and a reduction of the circuits of Berenice and Cyrene. Elsewhere, Justinian's rebuilding followed the Hellenistic line and the same defensive circuits continued until the Arab invasions of the 7th century, giving an overall life of some 900 years. Only one other city, Hadrianopolis, rose to importance in Roman times, but no trace of its walls survives today (Jones & Little 1971).

Ptolemaic Cyrenaica passed to Rome by the will of Ptolemy Apion in 96 BC and in 74 the first quaestor arrived. A period of peace followed which, although punctuated by the Jewish revolt of AD 115, lasted into the 3rd century. During this *Pax Romana*, however, some of the cities appear to have maintained their defences, repairing and altering walls, towers, and gates as necessary.

At Cyrene the acropolis gateway, the usual single portal flanked by projecting square towers, carried an inscription of Augustan date which stated that the walls of the citadel were repaired by the Proconsul Q Lucanius Proculus, and Goodchild's opinion was that most of the surviving masonry was either Hellenistic or Roman in date.

Berenice appears to have been the first city to have dismantled part of its walls, probably in the second quarter of the 1st century AD. Later rebuilding at Ptolemais may mean that some demolition was carried out there too. Rebuilding can also be seen between the east gate and the south-eastern angle at Arsinoë, where the wall is still standing to a good height. This consists of large drafted blocks laid in careful, regular courses, but of noticeably different workmanship from the earlier build. The west gate was also rebuilt though not at the same time. The rebuilding of a length of the west wall at Ptolemais has been dated by Kraeling (1962, 60–2) to the 2nd or possibly 3rd century on the strength of a reused Roman inscribed block. Either now, or a little earlier, the Barca gate was also walled off and closed, and alterations were carried out to the Tauchira Gate itself. Sometime after AD 250 a new Roman defensive wall was built at Berenice around a reduced area of the city (Fig 1; Lloyd nd, 165).

In AD 262 a crippling earthquake hit Cyrene, followed by another in 365, which also affected the other cities. At Apollonia, after years of neglect during protracted peace, the west gate (probably already dilapidated before the 'quake) was carefully repaired sometime during the next 50 years, as was the city wall. Both repairs were achieved by the reuse of whatever materials were available.

From AD 390 the nomadic Austuriani, moving on from Tripolitania (see p 17), increasingly laid waste the countryside and threatened the cities of the Pentapolis. The letters of Synesius, who became bishop of Ptolemais *c* 410, give us a very vivid picture of these unsettled times and the local defence forces raised against the desert raiders (Goodchild in Reynolds 1976, 239–54). The fortifications of Arsinoë and Apollonia were put back into repair but by now, if not earlier, the populations of Berenice, Cyrene, and Ptolemais had become so reduced that their circuits had to be drastically shortened.

Carthage and the Berber Kingdoms

In Africa generally, the defences of many settlements which later became Roman towns or cities originated in pre-Roman times. Although many early circuits were later replaced, some must have continued in use in early Roman times, while others—those at Lixus and Dugga for instance—continued to be used, in part at least, until late antiquity. Brief reference to pre-Augustan Africa is, therefore, useful.

It was the Phoenicians who brought the earliest outside influences into the Berber Kingdoms west of the Syrtic Gulf (Moscatti 1973, chs 7 & 9). By the time of the Punic wars (264–146 BC) the whole Mediterranean coast from Leptis to the Straits and the Atlantic shore as far south as Mogador (Essaouira), were dotted with settlements and colonies, while Punic influence, customs, and trade were reaching many inland sites. At their most sophisticated, as witnessed by Carthage herself and Lixus, the defences of these cities were well developed: at the opposite end of the scale, however, humble banks and simple ditches sufficed to protect others.

Carthage's triple line of defences was destroyed, like everything else, after the sack of 146.[1] But Thapsus, Acholla, Hadrumetum, and three other cities which joined Rome in the final round were created free *civitates*, while Utica received further grants and was chosen to be the capital of the province of Africa. Presumably each kept its existing fortification (certainly Utica, Hadrumetum, and Thapsus withstood sieges or were defended during the Civil War a century later), but no trace of these has to date been located, although Punic burials at Utica and a consideration of the site's topography give some idea of the settlement's probable extent (Lézine 1966; 1971).

To the west, over the half century 204–148 BC, Masinissa slowly built up a Numidian Kingdom stretching from the River Moulouya to the emporia of Tripolitania. Within this area the older Punic cities seem to have exercised a fair degree of autonomy. In the east, Leptis, Sabratha, and Oea (Tripoli) were flourishing ports. 3rd to 2nd century BC cemeteries at all three sites indicate areas lying outside their defences, but so far the defences themselves have not been located. In each case, as with cities elsewhere, later expansion removed the earlier fortifications, which were replaced by new circuits during the Roman period.[2]

The same state of affairs is found further west, where little of undeniably Punic date survives. Walling at Hippo Regius (Annaba), consisting of large, dressed blocks, some with rough bosses and drafted margins and occasional joggled joints, has previously been seen as part of the '*emporium primitif*'. Recently, however, this has been dated no earlier than 40 BC (Marec 1954, 16–17, 45, and fig 19; Lassus 1976, 394–6). Although ports like Igilgili, Tipasa, Gunugu, and Lixus were Phoenician in origin, their earliest surviving fortifications are of a later date. Lixus (Pliny, *Nat Hist*, 29.63; Euzennat 1976, 521), traditionally founded in the second millennium BC, has produced the most important surviving example of what is taken to be the sort of defence that the leading Punic cities were erecting in the century between the fall of Carthage and the Augustan colonization of Mauretania. The sophistication of the work is even more impressive for being located on the Atlantic coast of Morocco, well away from the main centre of Punic power (Fig 3), and where the Hellenistic–Mediterranean tradition of defences might not have been expected to be so strong.

The site crowns a prominent hilltop overlooking the River Loukkos a short distance from its mouth. Considerable stretches of wall survive surrounding the acropolis and the town in general. Excavation and aerial photography show it to have consisted of frequent, bold salients and re-entrants like the Hellenistic walls already described. Romanelli (1970, 80), who dated the earliest walls to the later 2nd century BC, described the eastern sector as constructed of huge, irregular, polygonal masonry, infilled with smaller rubble, side by side with regularly laid ashlar. The western sector was of massive, squared blocks of good quality stonework, varying in size from course to course and infilled with smaller material. Even to the casual observer, however, it is clear that there are many variations of construction, while sondages and random digging have produced lengths of high quality masonry, still *in situ*, at several points below buildings in the upper town, all clearly early defensive walls. The history of the site is considerably more complex than published accounts suggest and a proper study of the walls, together with selected excavation to date the various types of work, would be of the greatest value.

On the Mediterranean coast, the early defences at Tipasa (Gsell 1911, 4.38; Baradez 1954; 104–7), although undated, probably best fit into the same general period (Fig 1, wall A). There, Gsell recognized an early defensive circuit, which he called '*phénicien?*', enclosing no more than the Forum headland area of the later city. Baradez identified more of the same structure, which consists of a wall (and at least one tower) constructed of large, dressed facing stones of varying thicknesses, mortar-bonded. The narrow space between the wall faces was packed with rubble core. Baradez estimated the whole to have totalled 650 m of wall, with a further 600 m of steep cliffs (presumably the promontory itself) completing the circuit.

For the bulk of the cities of Numidia and Mauretania there is even less evidence, although a few examples may be quoted. Dugga (Poinssot 1958, 68), lying in a rich agricultural region between Carthage and the Numidian hinterland, was the seat of local princes by the end of the 3rd or early in the 2nd century BC; a little later it fell to Masinissa. A length of 130 m of early wall survives on the north-west side of the so-called pre-Roman citadel, the rest of the enceinte being classed '*basse époque*'. The stonework consists of two faces of rather rough stone blocks 2–5 m apart, carefully fitted together without mortar, with a rubble core between. The wall itself follows an irregular course towards the steep northern edge of a rocky scarp, and is reinforced by two projecting rectangular towers. No close dating is possible.

Siga, western capital of the Massaesylian King Syphax, remains mostly unexcavated and traces of walls on the site are as yet undated,[3] but its eastern counterpart Cirta (which was early lost to Masinissa) introduces us to an important class of site: the fortified spur or *éperon barré*. While some of these were given up in later prehistoric or early classical times others, including Cirta, have remained important cities to the present day. Little study of the type has previously been carried out (but see Gsell 1927, 232–45; Marion 1957). Found from Morocco to Tripolitania and down into the Sahara these sites are only now receiving fuller attention (eg Daniels 1968; 1972/3; Fentress 1979, 30–9; Jones & Barker 1979/80). Examples are noted below (p 13) from the region of Cirta and the plan of Magrusa in Wadi Zemzem, Tripolitania, is given as an example (Fig 3).[4]

Cirta (Ksantina, Constantine) (Gsell 1911, 17.126) occupies a strong natural position, a sheer-sided spur surrounded on three sides by the rocky valley of the Rhummel (ancient Ampsaga) and similar natural defences (Fig 1). Little in the way of man-made protection is needed on the one open side to make the site virtually impregnable, if resolutely held. In the century after 146 BC, Cirta became a key site in the Jugurthine War, first when Roman settlers helped to hold it against Jugurtha, then later when Metellus put his booty and prisoners into it for safekeeping; both events imply strong defences (Strabo 17.3.13; Livy 30.128.44; Sallust, *Jugurtha*, 21, 23, and 26).

Other native sites which were fortified during the Jugurthine War, or were at least defensible, include Lares (Henchir Lorbeus) (Ennabli 1976, 484), Vaga (Béja), Capsa (Gafsa), Sicca Veneria (Le Kef), and Thala, which was only taken after a 40 day siege by Metellus in 108 BC, while in 109 Zama Regia had withstood Metellus during an even more epic siege on account of its fortifications, which Vitruvius described as a double wall (Sallust, *Jugurtha;* Vitruvius 8.3.24). To these sites Dugga, Maktar, and Bulla Regia should probably be added as possessing or likely to have possessed defences.

Half a century later, Caesar's campaign in Africa against the senatorial forces and King Juba I of Numidia (47–46 BC) centred on Thapsus (Ras Dimasse). Cities mentioned as defended or fortified include Utica, Clupea (Kelibia), Thysdrus (El Djem), Hadrumetum (Sousse), and Thapsus itself, which withstood a major siege. Following his victory Caesar annexed Numidia, which he turned into a second province, Africa Nova, with its chief city at Zama. He also founded a new colony at Carthage in 44 BC and settled veterans at various other places including Clupea, Thysdrus, and Hadrumetum. Such settlements would almost certainly have been protected.

Augustan Africa

During the reign of Augustus the picture becomes a little clearer. When King Bocchus died in 33 BC Mauretania was annexed. Over the next eight years many Roman colonies were planted in Mauretania, Numidia, and the two Africas, and numerous other cities received some form of status. Pliny lists names (*Nat Hist*, 5.1.17–4.26), although mere lists tell us little; the remains of fortifications have been found at several sites including some not in Pliny's list. The importance of colonies has already been noted (p 5), with particular reference to such Augustan-date foundations in Narbonensis as Orange, Arles, Fréjus, Nîmes, and Vienne, each of which received an expansive circuit of walls and impressive gateways from the colony's foundation.[5] Italian parallels include Spello, Turin, and Aosta; while Autun, lying in Gallia Comata and not even a colony, is in some ways a more significant pointer. This Augustan foundation also received gates and lavish walls at its birth.

On the other hand, whether the defences noted by Gsell and others at Augustan sites can all be dated as early as Augustus is a difficult question. Procopius (*de Aedificiis*, 6.5) tells us that the Vandals destroyed the defences of many cities when they invaded Africa, only Carthage and a few others being spared. During the Byzantine occupation (AD 533–698) many city defences were rebuilt and additional fortifications were added (usually against the Berbers). This ought to mean that within the Vandalic-Byzantine area the majority of surviving walls are of 6th or 7th century date. Just how completely the earlier defences were levelled, however, remains a debatable point: early masonry clearly did survive, while the archaeologist has uncovered traces of circuits which had almost certainly been abandoned or were already lost to view by Vandal times. Some sites, however, wholly or mostly uninhabited in the early 19th century and covered with archaeological remains, now lie under thriving modern cities, while too often the wanton disregard shown to antiquities by the 19th century settlers has destroyed vital information, so

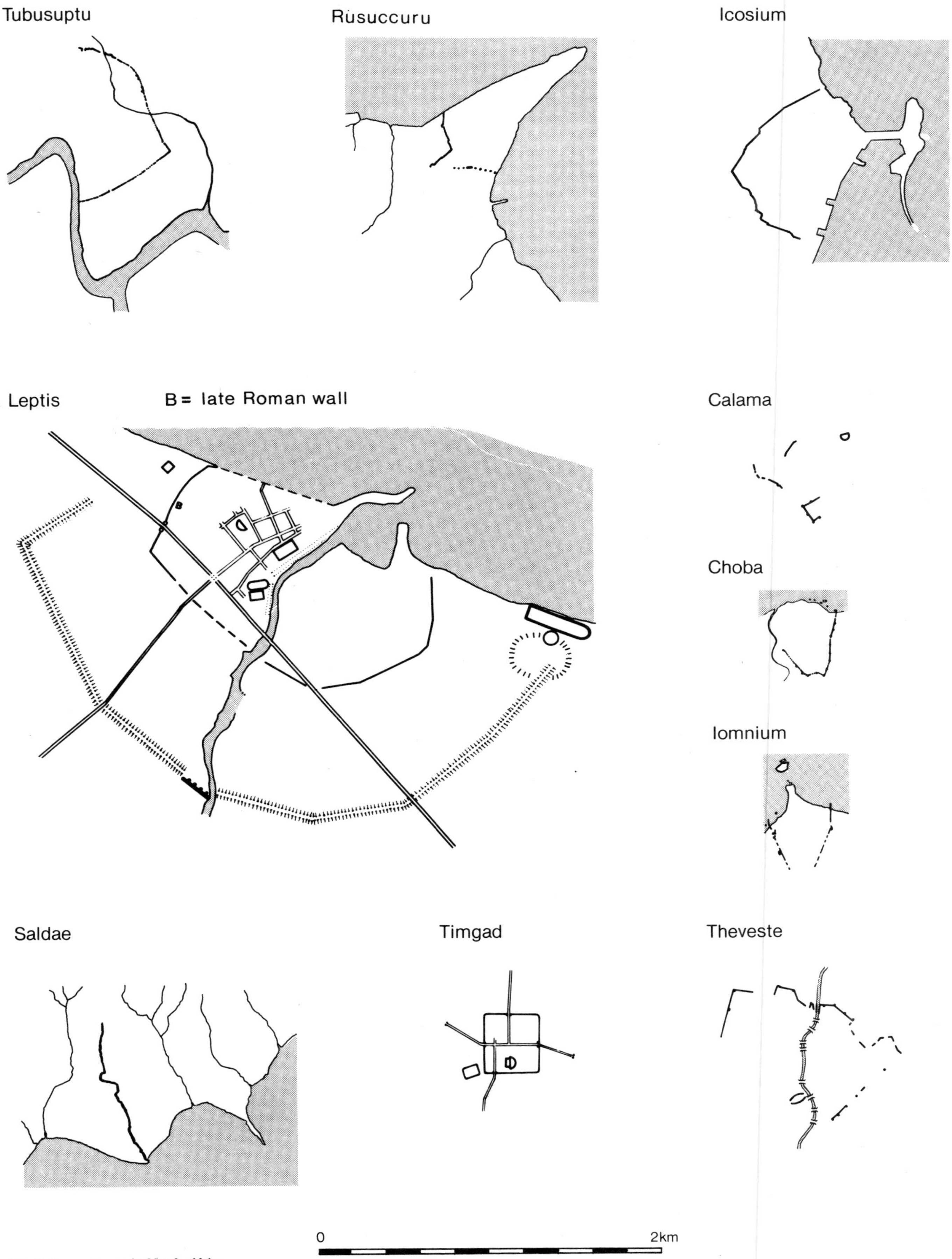

Fig 2 Roman towns in North Africa

that the brief accounts noted by Gsell and others are the only record of evidence now lost. This applies equally to post-Augustan sites.

The greatest density of colonies was in the two Africas, which in 27 BC were brought together with the military region of the south to become the single province of Proconsular Africa. Utica, Carthage, Thuburnica, and Vaga, *inter alia*, were already in existence and presumably possessed defences, likewise many of the free cities Pliny lists, including Hadrumetum, Thapsus, Thaenae (Henchir Thina), Tacape (Gabes), Oea, Sabratha, and Zama, most of which we have already met. The Numidian colonies, or towns, of Cirta, Sicca Veneria, Rusicada (Skikda), Bulla Regia, Hippo, *et al*, probably also possessed defences.

In Eastern Mauretania Rusguniae (Tamentfoust) (Fig 1), Saldae (Bejaïa) (Fig 2), Igilgili (Jijel), Tubusuptu (Tiklat) (Fig 2), Cartenna (Tenes), and Gunugu[6] have produced remains of ramparts. That at Gunugu is described by Gsell as part-built of reused material and therefore likely to be later, although Cagnat offers no date for it. Pringle (1981, 222–3) suggests that the wall (now destroyed) was the Byzantine city circuit although, as he admits, its date remains unknown. Even if it is late, it does not weaken the argument for early defences at this and the other Augustan colonies planted in Berber Mauretania. In the other cases descriptions are extremely brief: walls, where widths are given, vary from 1.10–1.50 m, and their construction is usually described as *en blocage*, ie rough stonework, core (Grenier 1931, 358 for the use of terms). In every case, however, it is only short lengths which have survived and these without dating evidence.

The colony of Aquae Calidae (Hammam Righa) may also be Augustan: it too has traces of a wall, with square towers.[7] Perhaps Rusuccuru (Dellys) and Iomnium (Tigzirt), discussed under Claudius (p 10), and even Icosium (Algiers), a Vespasianic colony (p 11), should be included in this group.

Banasa, on the banks of the river Sebou in western Mauretania, has been partly excavated. A length of something over 40 m of defensive wall has been uncovered, consisting of crudely faced stonework, laid in rough courses. This is usually dated late, but Euzennat has argued that it is more likely to belong to the Augustan colony (Euzennat 1976, 140–1; cf Rebuffat 1974, 512).

In 25 BC Augustus placed Juba II, son of Juba I, on the throne of Mauretania, returning the kingdom to native rule. It is to his long reign (25 BC–AD 23) that the early defences of his two capitals, Iol (Cherchell) and Volubilis, are usually attributed. The old Punic city of Iol (renamed Caesarea) was turned by Juba into a centre of Greco-Roman style and culture (Gsell 1911, 4.16; Duval 1946; Lassus 1976, 413–14). As well as imposing public buildings, it was provided with a truly Hellenistic enceinte *c* 7000 m in length (approx 4¼ miles), surrounding a little over 370 ha or 890 acres (Fig 1). The wall was built of rubble core set in mortar and faced with small ashlar blocks. Additions and later work are of a noticeably inferior quality, the facing stones often being extremely rough. Projecting towers occur at varying distances from 40/50 m up to 160 m apart, which, it has been argued, indicates the use of some form of artillery. Thirty-nine are known for certain, with another six probables. One is round, two gate towers are octagonal, and the rest are rectangular. The south gate is the only one to have been studied (Fig 4). It is deeply recessed, lying some 30 m back from the wall, and protected by two octagonal towers. The gate itself is a triple arch built before the walls, which ran up to its sides. At some date the east portal was closed by a wall, the foundations of which survive.

As at Ptolemais, the walls run inland for a considerable distance up on to the plateau behind the coast. Here, where they are most distant from the city centre, and therefore most vulnerable, three circular fortlets were built astride them, each 35–40 m in diameter. Two have square towers on their circumferences, fort 2 having four, two on the city side and two on the exterior.

At some date alterations and rebuilding were carried out, including the addition of stairways at the back of the wall beside most of the towers. The similarity of these stairs to those giving access to the rampart walk at Tipasa (p 15), where they are an original part of the rampart, suggests that they were an Antonine addition at Iol.

At contemporary Volubilis less is known about the defences, although more about the city itself (Euzennat 1957; 1976, 988–9; Luquet 1972: Jodin 1966; Boube 1967). The early site occupies the end of a low spur (called an *éperon barré* in one description) which drops to streams on its south and eastern sides (Fig 1, wall A). To the north traces of two tumuli have been found just beyond the line of the early walls, suggesting a cemetery there. Excavations have located some 350 m of the pre-Roman east wall and some 160 m of the north. Each is built in a series of salients and re-entrants akin to Lixus and the Hellenistic defences of the Pentapolis cities. The construction of the wall, as described by Jodin (1966), varies widely from large ashlar blocks, in places with joggled joints and sometimes with faces left rough except for their drafted edges, to mixtures of large and small stonework in random courses, sometimes even incorporating boulders and polygonal blocks. Jodin dates the beginning of urbanization at Volubilis to the middle of the 2nd century BC (1966, 218), but such a wide variety of stonework must raise some questions as to its contemporaneity—or even if it all be the same wall.

From Augustus to the Flavians

From the death of Augustus to the accession of Vespasian there seems to have been less attention paid to African city defences. Tacitus (*Ann*, 4.24) tells us that in AD 24 the rebel Tacfarinas besieged the town of Thubuscum (Gsell 1911, 18.297; Sherwin-White 1944, 9). Syme (1951, 113–15) suggests that this was, in fact, Thubursicu (Thubursicum Numidarum or Khamissa), which became a *municipium* under Trajan. The siege was quickly raised, but the city must have had defences, although no part of them survives today.

Caligula had King Ptolemy, successor to Juba II, assassinated, and Mauretania rose in revolt. Several cities are believed to have been sacked at the time, Lixus, Tamuda, and Banasa amongst them, but Volubilis remained loyal and intact, and was handsomely rewarded as a result (*CAH*, **10,** 652). On restoring peace, Claudius divided the old kingdom into the two provinces of Mauretania Tingitana and Caesariensis, each named after its principal city. The continuing necessity of defences, in the west at least, must surely have meant that particular attention was given to fortification when sites were rebuilt. Colonies were planted at Tingi (Tangier) and Lixus, while Volubilis became a *municipium.* Of the old Punic city Tingi we know virtually nothing and Lixus already possessed defences, as we have seen. Volubilis was probably still within the early defences which had protected it during the uprising.

Further east, the walls of Iol may have been strengthened. A colony was established there with others at Rusuccuru, Oppidum Novum, and Tipasa. Rusuccuru has previously been identified with Tigzirt (Gsell 1911,

6.24, 34, 87) but Février (1976, 777) argues for Dellys, which is followed here. A considerable length of wall core (*en blocage*) was noted by Gsell (1911, 6.24), including an angle tower with a frontage of 7 m (Fig 2). There is no indication of date.

Tigzirt (Gsell 1911, 6.34), now equated with Iomnium, should perhaps also be noted (Fig 2). It began as a native town but is called a *municipium* in the Antonine Itinerary. Here too, Gsell recorded a wall *en blocage*, with what looks like a west gate defended by round towers and traces of another round tower (or gate tower) on the east side, all now almost totally destroyed. In view of the round towers an earlier date would fit this system better than a late one.[8] The site of Oppidum Novum is known, in the Chélif Valley, but little else about it (Leveau 1972).

The Roman-date wall of Tipasa is usually dated to Antonine times, although Euzennat has recently written (1976, 925) '. . . in AD 40. At this time . . . a vast enceinte was built. It enclosed some of the ancient necropoleis (especially those to the E), which were covered by the expanding settlement'. In some ways an early date for the start of the new city wall would fit better than a later one (see below).

During the civil war which followed Nero's death Leptis and Oea literally came to blows with one another. Finding herself the loser, Oea called in the wild tribe of Garamantes to ravage the territory of Leptis, 'so that the inhabitants cowered in fear behind their walls' (Tacitus, *Hist*, 4.50). The only defence known which could fit this date, the large 'Monticelli', is an earthen bank and not an actual wall (Goodchild & Ward-Perkins 1953; Romanelli 1970, 82). Here, however, is a work on the grand scale, some 5500 m in length and enclosing 425 ha or 1050 acres (Fig 2). It has been suggested that it may have been a hastily constructed protection against the raid, the city having long outgrown its earlier fortifications, but it encloses a far greater area than the city alone and it would surely have taken some time to build.

The Flavians and Trajan

Under the Flavian emperors and Trajan there was a spate of over twenty new colonies and *municipia* in Africa (Gascou 1972; Leglay 1963; Haywood 1938, 37–9). Many were existing cities elevated, two resulted from the moving of Legio III Augusta, and the rest were new creations. Hippo Regius, Bulla Regia, Thubursicu, Capsa, Hadrumetum, and Leptis were all well established cities which were elevated.

Icosium (Algiers) was of Punic origin but did not become a colony until the reign of Vespasian (Pliny, *Nat Hist*, 5.1.20; Gsell 1911, 5.11; Le Glay 1968). Gsell noted the remains of Roman work in the old Turkish walls and elsewhere, and for once the description is comparatively full (Fig 2). One length of 100–150 m consisted of facing stones set in mortar, carefully constructed, with a square projection which could have been a tower. At two other points the remains of round towers survived and at a third a further length of faced wall, of excellent workmanship. Finally, between this last length and one of the towers there was another piece, *en blocage*. As suggested previously of Tigzirt, where circular towers also occurred, on average an early date probably fits these defences better than a late one.

Towards the Numidian frontier lies Sétif (ancient Sitifis), the only known colony of Nerva (Gsell 1911, 16.364; Février *et al* 1970, 70–86; Février 1976, 844). Again, the settlement was one of considerable antiquity and was fortified from an early date. Gsell noted that two circuits of walls once surrounded the city, but recent work has shown that both are late (one Byzantine). The early defences still await discovery.

Nearby, but beyond the River Ampsaga and therefore in Numidia, a new foundation was planted by Nerva at Cuicul (Djemila) (Gsell 1911, 16.233; Allais 1971; Février 1964, 4–26; 1978; Romanelli 1970, 85–6), which has been extensively excavated and well studied (Fig 3). Much of the line of the original wall has survived. The site sits towards the end of a steep spur and is polygonal in shape, 400 × 200 m at maximum and some 5·5 ha, 13 acres, in size, with a principal street running along the ridge of the spur from the south gate to the north. The city wall is 2–3 m wide, laid on a foundation of large blocks, and consists of two faces of medium-sized stonework with a core (*blocage*) of boulders and cut stone. The south gate is a single portal with slightly projecting towers, built of massive masonry with large quoins. Its passageway stonework has random bosses, and consists of well cut uncoursed ashlar. Later, the colony outgrew its original defences and reached some 26 acres in size.

Further east again, within the region which had resisted Tacfarinas, but (like Sétif and Cuicul) in countryside which was still under-urbanized, Thubursicu Numidarum (Khemissa) was created a *municipum*.[9] At the same time a veteran colony was placed at Madauros (Mdaourouch, Madaure) (Gsell 1911, 18.432) as, it has been suggested, 'it evidently was considered prudent to have certain centres of reliable veterans among the strongly native population' (Haywood 1938, 38). In such a climate strong defences would be essential, and the same holds true for nearby Calama (Guelma), another older settlement which also became a *municipium* (Gsell 1911, 9.146; Gascou 1972, 143). There Gsell recorded remains, since destroyed, of a Roman enceinte to the north-west of the Byzantine fortress (Fig 2). He also pointed out that the west, south, and part of the north side of the fortress itself were of superior quality to the rest and likely to have been of earlier date—in fact, a part of the city defences. The date is quite uncertain, and Pringle (1981, 188–90 and fig 24) thinks the whole fortress late.

South, in the more open country seasonally grazed by the Gaetulian tribes, especially the hostile Musulamii, Sufetula (Sbeïtla), Cillium (Kasserine), and Thelepte have all been argued to have started life as forts (Haywood 1938, 37, following Broughton 1929, 95). At each a settlement must quickly have taken root, for each is believed to have become a *municipium* under Trajan, as did the old established settlement of Capsa (Gafsa) in the same general area. A little to the west, but still on Gaetulian land, Ammaedara (Haidra) and Theveste (Tébessa) were both made veteran colonies on successive movements of the Legion west (Fig 2). Nothing is known of the early stage of either; each grew to some size, but their legionary ramparts can be expected to have provided protection during their early years (Ennabli 1976, 50; Gsell 1911, 29.101; Février 1976, 913–14).

Tacitus, in his description of Tacfarinas' revolt, speaks of the Musulamii as bordering the empty wildernesses of Africa which were even then without urban life (*nullo etiam tum urbium cultu*; Tacitus, *Ann*, 2.52). The Flavian-Trajanic policy of settling veterans throughout this southern, townless area was also carried far to the west. Theveste must be seen as the easternmost of a line of settlements running along the northern foothills of the Nementcha-Aurès range. To the west lay Mascula, now covered by modern Khenchela but probably a veteran settlement of some sort, and then Timgad, ancient Thamugadi (Gsell 1911, 27.255) This colony was deducted

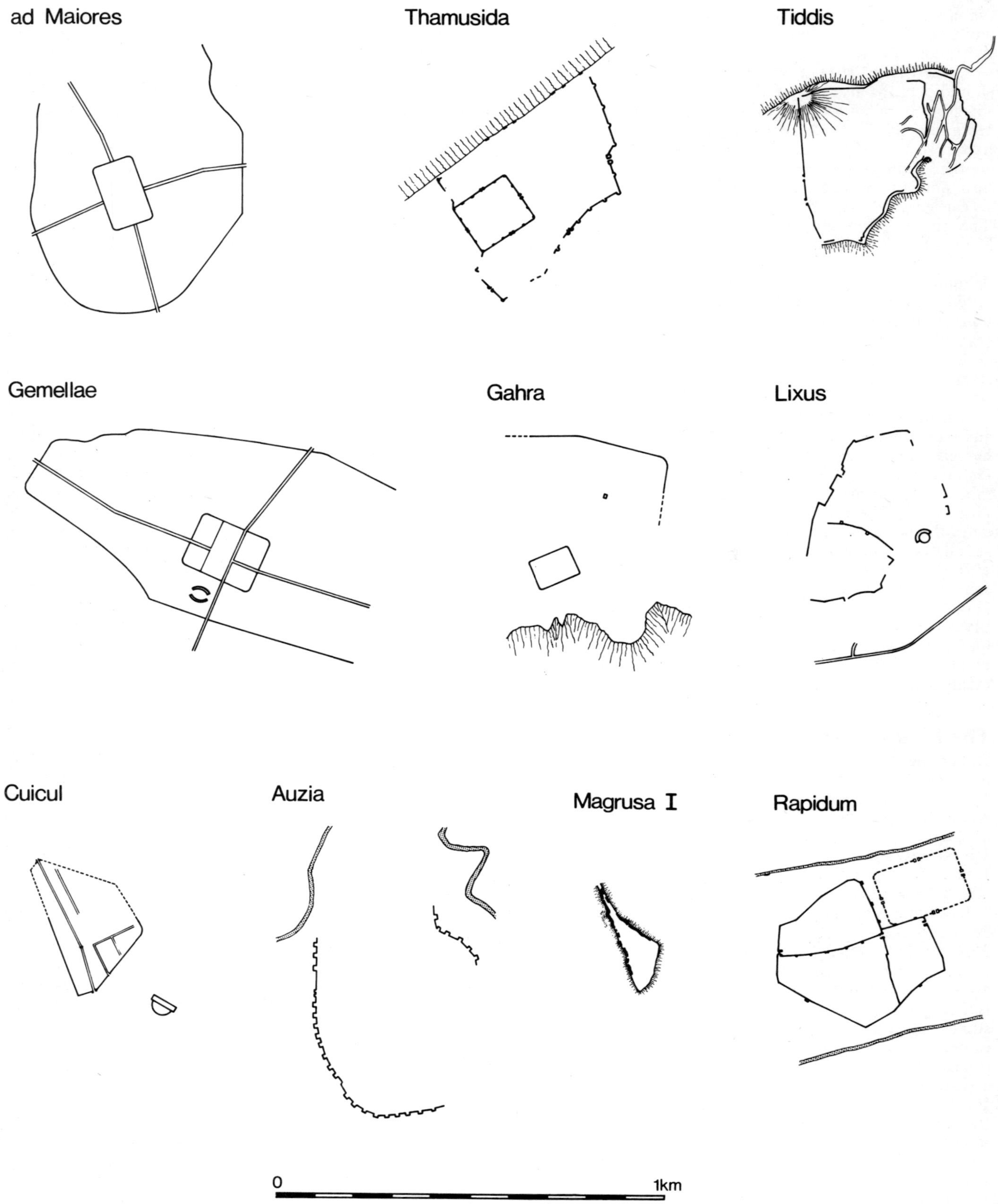

Fig 3 Roman towns in North Africa

by the legate Municius Gallus in AD 100 and built by the Legion. Thanks to large-scale excavation it is one of the best known and studied sites in Roman Africa. The original settlement was *c* 355 m square and enclosed by a wall with gates, rather like a legionary fortress but only 12·6 ha, 30·25 acres, in size (Fig 2; Lassus 1969).

The wall appears, where it now survives, to have been of rough workmanship. Ballu (1903, 5–15), who even thought it was a late work on the line of the early defences, described it as made of large-sized, roughly-worked facing stones bonded with earth, except on the north side, which was in places of a little better quality. Others, also noting the unexpectedly rough quality of the work, have thought the wall belonged to a military fort placed here before the colony was created. Certainly, where it survives today, the work appears very rough, but two points argue against an initial fortress. First, a small hillock was purposely included within the area of the colony, in order to provide a natural site for the theatre: within a legionary fortress such an inclusion would have been a considerable inconvenience. Second, the south gate of the colony is not where the legionary *porta decumana* should have been, nor could a normal *porta decumana* have been built, yet the surveyors could easily have chosen a site on flat ground where a regular fortress plan could have been achieved.

The surviving lower part of the north gate is a single portal constructed of monumental ashlar (Fig 5). No doubt begun in AD 100 (like the original east gate) it was not completed until the reign of Pius.

The Confederation of the IIII Coloniae and Hadrian

Cirta, the old Numidian capital, had been granted by Caesar to P Sittius, a companion of his and a soldier of fortune. Created a colony at an early date, it was part of Proconsular Africa until the reign of Hadrian, or perhaps Trajan, when 'one of those anomalies common to the administrative history of Africa occurred'.[10] This was the formation of the Confederation of the IIII Coloniae, which Hadrian removed from the Proconsul of Africa and placed under the legate of Legio III Augusta. Cirta was the principal city, the other three being the colonies of Rusicada (Skikda), Chullu (Collo), and Milev, each newly elevated (Gsell 1911, 8.196; 8.29; 17.59). Cirta, alone, has produced traces of defences which, as already said, began in pre-Roman times and continued in use until long after the fall of Roman Africa (Fig 1). Gsell recorded remains of walling in monumental stonework across the neck of the plateau, of which one square projecting tower constructed of large masonry survived until 1867.[11] A further length of wall, with a semicircular tower, was visible on the west side of the city. Whether any of these dated to the High Empire is uncertain: other lengths were almost certainly later.

Pringle (1981, 194–5) in fact assumes all the recorded walls to have been of late Roman or Byzantine date. However, Cirta was a place of major importance and one we know to have been strongly defended from an early date (p 8). It is not unreasonable to expect its defences to have been commensurate with those already described at Lixus, Volubilis, and Tipasa (which in part at least were constructed of large-sized ashlar) and lengths of the earlier walls to have been incorporated in the later circuit.

Cirta also possessed *castella*, distributed throughout its territory, the names of several of which we know (Février 1976, 205, 224–5).[12] Tiddis (Castellum Tidditanorum) occupied a rocky hillside, with parts of its strong, pre-Roman defensive circuit surviving (Fig 3), including a gateway built of large, irregular stones. In late Roman times the upper part of the town and the plateau above it were surrounded with a new defensive wall, built of reused material, mostly large blocks, but here and there reusing stones from the earlier fortifications (Gsell 1911, 17.89; Berthier 1951). Here, a pre-Roman, native hill-town rose in status and importance and continued to flourish after the dissolution of the Confederation. Its most famous citizen was Lollius Urbicus, who achieved the consulship under Hadrian, reconquered Scotland for Pius, and was given the prefectship of Rome (the pinnacle of a senatorial career).

At Thibilis (Tibilis, modern Announa) another Cirtean dependency grew into a town of *c* 25 ha, 60 acres, and became a *municipium* in the 4th century (Gsell 1911, 18–107). Situated in mountainous country these last two sites fall into the same category as Cirta itself, that is, fortified spurs with steep or rocky sides (see p 8). Others to the west, near Sétif, include Henchir el Ksar (Gsell 1911, 16.196), a rocky promontory which has produced an inscription of AD 157 mentioning an *ordo*, while Cuicul itself is a not dissimilar site (Fig 3).

Hadrianic colonies and *municipia* lie mostly in the rearward areas of Proconsular Africa and came about by the elevation of long established cities such as Utica, Lares (Henchir Lorbeus), Bulla Regia, Zama, and Althiburos (Ebba Ksur). Thaenae (Thyna, Henchir Thina), some 7½ miles from Sfax on the Tunisian coast, also became a colony (Ennabli 1976, 898–9). The city walls have been located. Ennabli talks of a monumental enclosure of which one side was 2·5 km long, including a gate with two semicircular towers. Another source is more specific: 'it is surrounded by a vast line of walls, dating from the second century, with round towers at regular intervals and monumental gateways' (North Africa, 103, entry unsigned).

Choba (Ziama), on the Mauretanian coast, became a *municipium* (Fig 2). Gsell (1901, 99; 1911, 7.68) records its walls as built of a core *en blocage* with two faces of small, regularly cut stones, well laid. Internally, the wall was buttressed at intervals of 2·30 m. These buttresses carried arcades supporting the rampart walk. Numerous protecting towers were 30–50 m apart, with a single large circular tower 7 m in diameter at the highest point, on the south side. No certain trace of stairways survives, and the gateways cannot now be identified. Pringle (1981, 256) briefly dismisses Diehl's late dating of the town defences, while proposing a Byzantine date for the late reducing wall (at the same time he admits that it could have been of 4th or 5th century date). The main circuit, as that at Thenae, could have been built at almost any date.

The Antonine age

Pausanias (8.43.3) and the Augustan History (Pius, 5.4) tell us that the Moors took up arms against Pius, who, through legates or governors, forced them to sue for peace, having driven them as far as the Atlas Mountains. Did such a war produce a spate of city defences? The older view saw Africa aflame from the Aurès to the Atlantic, with a spate of defences, from Tipasa and Sala in the 140s to Volubilis and Rapidum in the 160s, indicating continued insecurity.[13] Recently, however, Rebuffat and Février have reached rather different conclusions (Rebuffat 1974, with refs), in which they reduce the military activity and see city walls as much as indications of prestige and civic dignity as of defence. Of the two views this seems the more acceptable, for at the very moment when Volubilis and Rapidum were fortifying, Timgad, on the very edge of the supposedly hostile Aurès, was removing its walls and

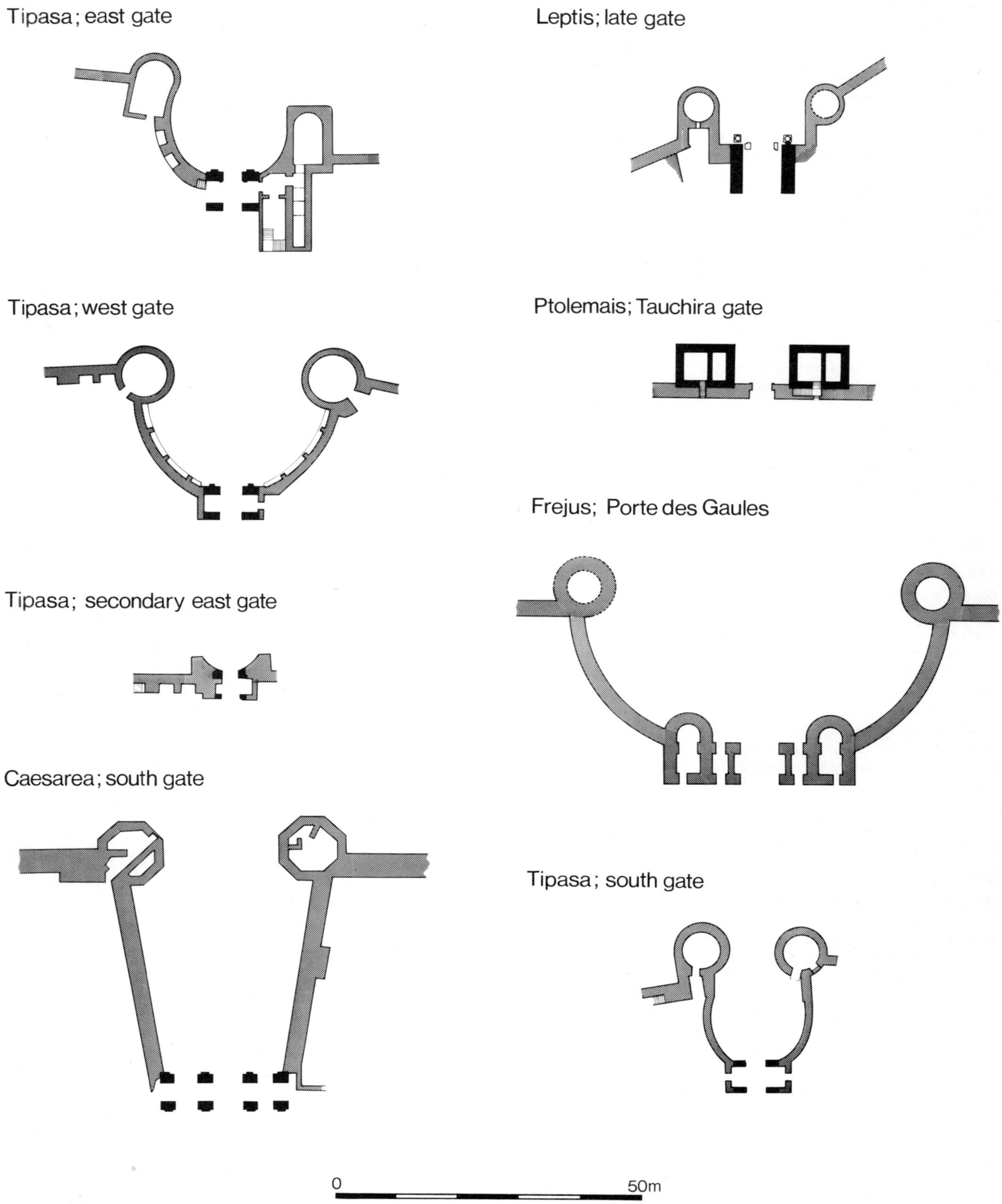

Fig 4 North Africa: Gates

Cuicul, in the foothills of the Petite Kabylie, was spreading beyond its defences.

Two cities, however, did receive new defences under Pius. At Sala (Rabat), a long inscription of AD 144 records that Sulpicius Felix, prefect of Ala II Syrorum, had the *municipium* surrounded where it was most vulnerable by the strongest of walls, and at the lowest of costs (Carcopino 1943, 200–28; Rebuffat 1974, 501–6).

Tipasa (Gsell 1901, 98–9; 1911, 4.38; Duval 1946; Baradez 1952; 1966a; cf Grenier 1931, 294–314) also received new walls (Fig 1). As said above, Février suggested that these were Claudian, but a fragmentary inscription from the minor east gate dates to AD 147. Unless this is an added feature, it must date the whole enceinte as well as its gates. The city wall, some 2300 m long, is *c* 1·60 m wide and built of a rubble and mortar core, faced with small-sized masonry. Twenty-seven definite towers and perhaps 32 in total projected from the walls, 11 (or 18) round and 14 (or 17) square. Two large round towers had internal staircases; elsewhere they were external and give an estimated height of *c* 6·70 m for the rampart walk, to which another 2 m should be added for parapet and crenellations.

The three main gates were of a particular form (Fig 4). Each was recessed back from the walls by the use of a semicircular forecourt protected by projecting guard towers. The west gate is reminiscent of those at Fréjus (Fig 4), Arles, and Aix in plan (if somewhat simplified) with a wide, sweeping semicircle in front of the actual gate, the walls of which were decorated with pilasters. The south gate was a more sophisticated adaptation with only a narrow passageway between the guard towers. The gate itself was a single portal of large-size ashlar. The east gate had basically the same arrangement, but the minor east gate was a much less spectacular affair (Fig 4) with slightly curved cutaway faces to its guard towers, foreshadowing the principal gates at the Severan outpost forts of Bu Ngem and Gheriat.[14] In passing, however, it should be noted that the workmanship of the walls and gates at Tipasa shows considerable variations, as Duval's photographs make clear, and a further study of the whole might be rewarding.

About the same time that these defences were being constructed the north (AD 149) and east gates (AD 146) of Timgad were completed in an ornate, monumental style (Fig 5), each being decorated with engaged pillars. This fashion of decorating gates more and more in the manner of honorary arches became increasingly common in Africa during the 2nd century.

Twenty years later Rapidum (Sour Djoub) and Volubilis received walls. The *vicus* at Rapidum (Fig 3) had grown to some 15 ha, 36 acres, and under Marcus and Verus (AD 167) it was provided with a wall (Gsell 1901, 91–6; 1911, 14.90; Seston 1928 (plan not accurate); Rebuffat 1974, 521; *CIL* 20834–5). Later, this was rebuilt, but the large, conspicuous, good-quality ashlar is probably part of the early wall.

A year later Volubilis received new walls (Euzennat 1976, 989; Rebuffat 1974, 510–12; *World Guides* Morocco, 222). These run for 2350 m, or just under 1½ miles, enclosing an area of *c* 40 ha, 100 acres, which included the newly constructed north-eastern quarter (Fig 1). The walls were built of a mortar-and-rubble core with small ashlar facing stones. Eight gates are thought to have given access. Several were single portals, but the north-western and Tangier gates had side passages as well as the main carriageway and the south-western was double (Fig 5). These gates were built of large, monumental ashlar, while their towers and the interval towers (of which there were 35–40, all projecting) were of much smaller stonework.

Study of the Antonine walls shows that the original scheme had been to enclose a limited area, entirely on the west side of the Oued Fertassa, with the new walls running to join the older enceinte at its north-eastern corner. A length of this wall was built but was abandoned when it was decided to take the circuit across the Oued and include the so-called Temple B.

A few other smaller sites in Tingitana received defences about this time. At Thamusida, for instance, a fort of 2·3 ha, 5·5 acres, was built on the south-eastern side of the town some time after AD 166 (Euzennat 1976, 902; Rebuffat 1974, 506–10). Later, under Commodus, the town was surrounded by a wall which abutted the fort on either side (Fig 3). Three gates with horseshoe-shaped towers and thirteen interval towers have been located.

In western Mauretania, while Volubilis and Rapidum were receiving defences, Timgad (Gsell 1911, 27.255; Lassus 1966; 1969; 1976, 899–902; Courtois 1951) was clearly spreading beyond hers and not replacing them, a significant event for a major city on the edge of the mountainous Aurès.[15] A little over 200 m beyond the original east gate a new gateway was built, matched by another 340 m along the western street (Fig 5). Each was an arch-like structure and were it not for pivot-holes could easily have been taken for a freestanding monument. The western is dated to AD 167, but the two are definitely a pair.

In African cities the distinction between arches and gates becomes increasingly hard to make as the 2nd century progresses, with gates losing their defensive nature and becoming totally freestanding structures.[16] At Thibilis (Annona) a gate has attendant city walls, but one at Lambaesis (Fig 5), built in the reign of Commodus by Valerius Maximianus, seems to lack walls, while further structures at Lambaesis, Verecunda, Diana Veteranorum (Fig 5), and elsewhere are clearly built as free-standing arches. At Timgad, the old west gate was replaced by a new, monumental arch (Fig 5). In what must have been seen as a time of overall peace, gates clearly became solely monuments of civic pride and dignity.

Severus and the 3rd century

Lambaesis (Janon 1973; Gsell 1911, 27.222–4) had become a *municipium* under Marcus; under Severus it was made the capital of Numidia and (now or under Caracalla) raised to the status of *colonia*. So far no city walls have been located, although arches and arch-like gateways have been found in both the city and the legionary *canabae*.

Under Severus the area of the Numidian military command was dramatically increased by the construction of outpost forts in the Tripolitanian pre-desert and an advance of some 200 kms through the Zab and Ouled Naïl hill-country. Some of the more advanced posts (Castellum Dimmidi, for instance) were abandoned within about a quarter of a century, but others continued to be held for very much longer and sustained sizeable civil settlements (Picard nd). One such is Gahra (Gsell 1911, 47.1), where a small fort of *c* 1 ha (2–2½ acres) was surrounded by a settlement of *c* 20 ha (40–50 acres). Gsell's informant recorded walls 1·50 m wide, flanked by bastions and pierced by four gates (one per side), traces of internal rectilinear streets, and a further north-eastern enceinte, none of which is particularly clear today after intensive agriculture (Fig 3). In addition to military inscriptions the site has produced tombstones of children in their teens and women in their thirties. The latest coins recorded date to the reign of Constantius II (337–61), while the fine wares

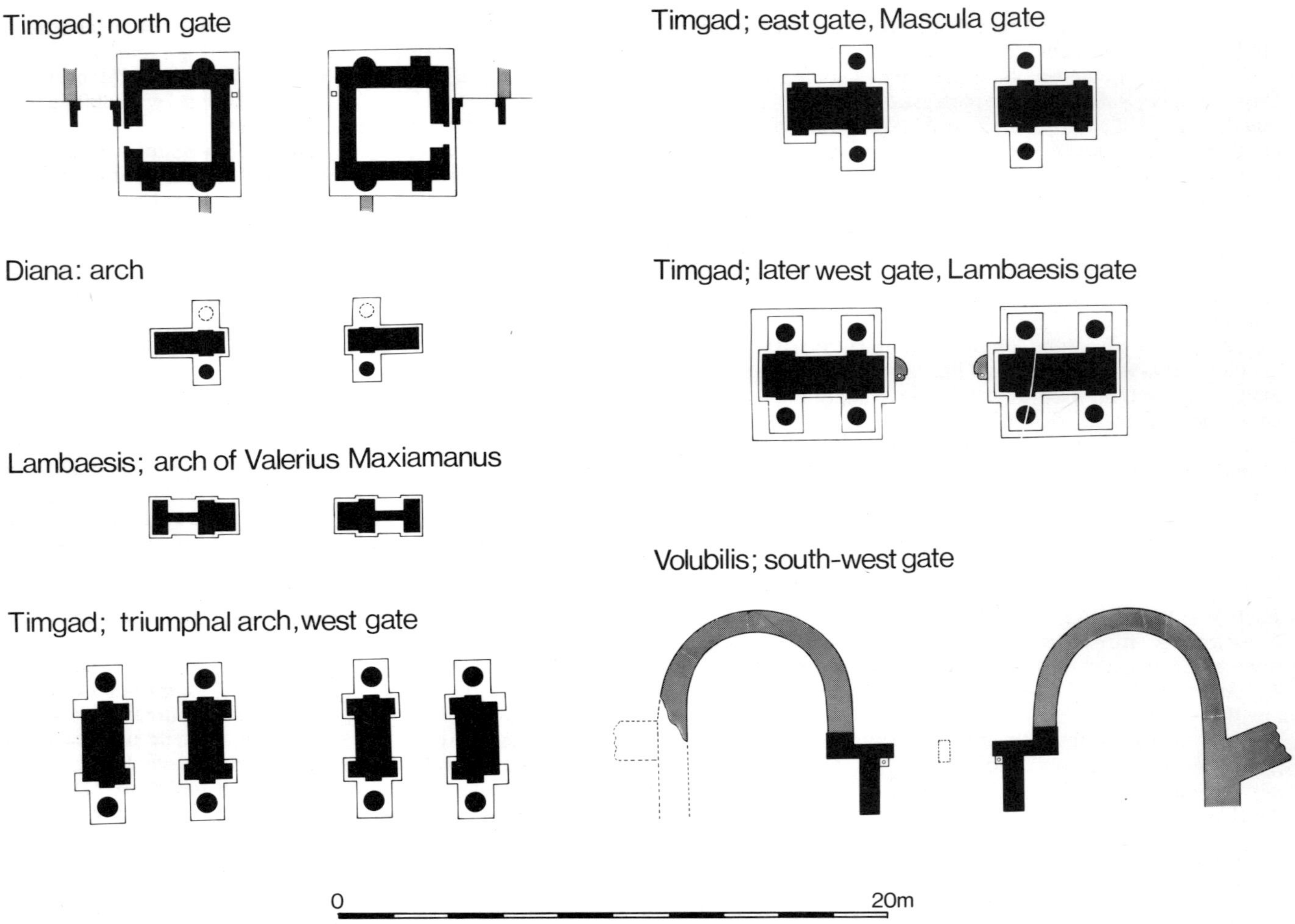

Fig 5 North Africa: Gates and arches

run on into the 5th century (*CIL* 8784–90; Carcopino 1925, 145).

To the east, along the older sector of the frontier, similarly long-lived settlements are known. At Gemellae, a Hadrianic foundation, a fort of nearly 3 ha, 7 acres, is surrounded by a large civil settlement called, on an inscription of AD 257–8, *municipium Gemellensium* (Gsell 1911, 48.39; Baradez 1949, 100–6; 1966b; Trousset 1976). The whole is enclosed by a bank or wall 2800 m in length (Fig 3). A vast fortified site lies 6 km east of Ouled Djellal in Oued Djedi, as can be seen in a published photograph of Baradez (1949, 124; presumably Gsell 1911, 48–79). No study of surface material has apparently been carried out, but the site looks as if it had a long occupation.

Of the settlements at Biskra (Gsell 1911, 48.9) and Thabudeos (Thouda) (Baradez 1949, 282; Gsell 1911, 49.1) little is known, but that at Badias (Badès) is called a *municipium* on a fragmentary 3rd century inscription. Pieces of the enceinte wall, with round towers of large stones, are recorded by Gsell (1911, 49–51), although whether these surround the fort or the settlement is uncertain. East again, at Ad Maiores (Henchir Besserrani), a settlement of considerable size is called a *municipium* on an early Diocletianic inscription (Gsell 1911, 50.152; Baradez 1949, 118a). It contained two arches, one of which perhaps became a gate, in an enceinte 1800 m long and flanked by numerous towers (Fig 3). Said to be Byzantine, it was probably begun much earlier.

In Tripolitania, a *vicus* of 18 ha, 33 acres, lies on the north and east sides of the fort of Bu Ngem. Recent work has produced an encircling wall or walls, crudely built of faced stones set in coarse mortar. Its excavator pointed out that this was certainly no military defence against assault, but should be seen in terms of policing and property protection (Rebuffat 1969/70, 21 and fig 1). Whether similar fortified *vici* occurred at Gheriat el Gharbia and Ghadames is not clear; certainly civilian tombstones have been found at each site.[17]

Many years ago Sherwin-White (1944, 9) noted 'The significant thing in Roman Africa is the continuous development of the large village into the small town: the *castellum* becomes a *civitas* or *municipium*.' In Caesariensis we have already seen *castella* in the region of Cirta. South of Sétif, in the plains which stretch to the Hodna, a group of settlements on Imperial estates, also known as *castella*, shows growth in the early 3rd century. Many of these increase in size and acquire encircling walls, although whether for purely defensive reasons or partly as indications of their increased status is to a certain extent disputed (Matthews 1976, 164–5, with full refs to Février's views). Kherbet Aïn Soltane, Castellum Citofactense (Gsell 1911, 16.391), Kherbet Ksar Tir, Castellum Vanarzanense (Gsell 1911, 16.360), Guellal, Castellum Dianense (Gsell 1911, 16.368), and Kherbet Zerga, Castellum Cellense (Gsell 1911, 26.135), have each produced inscriptions of Severus Alexander or Gordian III recording walls, while actual traces of walls have been noted at Kherbet Aïn Soltane.

To the west of Sétif at Sertei, Kherbet Guidra (Gsell 1911, 16.34), a site 55 ha (132 acres) in size, has also produced an inscription of Severus Alexander which records a wall being built by its inhabitants. Traces survive; while at Satafis (Gsell 1911, 16.177) there are remnants of two ramparts with bastions.

Elsewhere in Caesariensis a new frontier (*Nova Praetentura*) was created by the Severan advance beyond the Titteri-Bibans and Ouarsensis ranges (Salama 1953/5). The older *vici*, if they had not already done so, now developed into settlements with the departure of the army. Both these and the *vici* of the new forts are dealt with in the next section.

From the death of Severus Alexander Africa faced troubles, both political and military, which cannot have been helped by the disbanding of Legio III Augusta in AD 238. But in spite of this there was no real breakdown during the next twenty years either in Numidia or Mauretania.[18] On the other hand walls, which had begun as the essential defensive requirement—later to become increasingly prestige items, if they were provided at all—now returned as a necessity.

From the Tetrarchy to the Vandals

The new order which was introduced after the civil wars and barbarian invasions of the 3rd century was dramatically different from that which had gone before it. Defensive systems were now no longer limited to the frontier zones of provinces, but were constructed over the entire countryside by means of a network of fortified strong points which served both as bases for troops and refuges for civilians. This may be seen in Africa, as elsewhere, with old forts restored, new ones built, and city defences brought up to date.

Ammianus records the raiding of Tripolitania by the Austuriani in the 360s and Goodchild and Ward-Perkins (1953) suggest that Leptis received a new circuit, enclosing 130 ha, 325 acres (Fig 2, wall B), at some date between AD 250 and 365. The two have been seen as related. The new wall consisted of a rubble core faced mostly with large-size masonry. It incorporated a 2nd century arch which was turned into a late version of the forecourt type of gateway (already seen at Tipasa and Caesarea) by the addition of a pair of projecting towers (Fig 4). The arch itself formed a recessed, single-carriage gate with a rectangular-shaped forecourt. As yet no matching eastern or southern gateway has been located.

At Sabratha a length of late wall has been found at the eastern end of the excavated area. Its precise length is unclear; Ward-Perkins (1976, 779–80) and Haynes (1965, 127) each suggest only a shortish fragment, but another published plan shows a more sizeable length, together with a gate (North Africa, 89–90, entry unsigned). Archaeological evidence suggests that Sabratha was sacked in the period 363–5, although Ammianus (28.6; cf Haynes 1965, 58) omits all mention of the city from his account of the Austurianian raids. Leptis and Oea fared better, although each had its territory devastated and Leptis actually withstood an eight-day siege.

Ammianus' account of the revolt of Firmus (AD 372–4) includes the names of various cities (29.5). Of these Tipasa withstood Firmus, but Caesarea was sacked (no doubt its huge circuit proved too long to hold) and Icosium surrendered, or was taken (Figs 1 and 2). The walls of Tubusuptu have already been noted (Fig 2) and an inscription tells us that *horrea* were built there in AD 304. At Auzia, certainly a *municipium* by the end of the 2nd century (Tacitus, *Ann*, 4.25; *CIL* 9046), Gsell noted that at the start of the French occupation (1843) lengths of rampart were visible, built of large ashlar blocks without mortar (Fig 3). This wall, which had frequent bastions, enclosed an area of at least 300–400 m by *c* 800 in size (Gsell 1911, 14.105). No date is given.

Parts of the wall of Zucchabar survived and are recorded as having been *en blocage* and of brick, suggesting that they had been rebuilt or repaired at a late date (Gsell 1911, 13.70). Another site which has produced evidence of defences is Mouzaïa, where an inscription records that the walls were restored under Constantius II and traces of an enceinte once visible included a monumental gate with two bastions (Gsell 1911, 14.1; *CIL* 9282).

Provincial reforms early in the reign of Diocletian created Sétif the capital of a new province: Mauretania Sitifensis. The main period of new building, however, came not at once but in the second half of the 4th century, particularly in the reign of Valentinian. Slightly earlier, between 355 and 378, a huge new rampart of over 5 km in length was built around the north-western quarter of the city, where many older buildings were restored and new ones erected. Workmanship varied from sector to sector, and reused material was incorporated, but most of the surviving masonry is good, dressed ashlar or hammer-dressed blocks with drafted margins (Février 1964, 26–33; 1976, 844–5; Février *et al* 1970).

Several sites just mentioned lie on the old Hadrianic frontier, abandoned when the Severan advance occurred. Others include Thanaramusa Castra (Berrouaghia) where some 8 ha, nearly 20 acres, of ruins are surrounded by an enceinte of roughly rectangular plan (Gsell 1911, 14.58). Likewise, at Lambdia (Médéa) traces of walls have been recorded (Gsell 1911, 14.48) as at Mina (Relizane) and Castra Nova (Mohammadia) (Gsell 1911, 21.36 and 27).

To the east both Thamallula (Ras el Oued) and Thubunae (Tobna) are known for their Byzantine defences. The *municipium* of Thamallula (Aïn Toumella) (Gsell 1911, 26.19), however, has a second defensive wall, which could have been late Roman in date, while Baradez's published aerial photograph of Thubunae (1949, 293; Gsell 1911, 37.10) shows two circuits of defences, one within the other, as well as the detached fort. Many of these settlements had long lives and the precise date of their defences is uncertain.

Finally, we have the Severan frontier line. Here, the *Nova Praetentura* created on the edge of the high plains south of the Tell Ranges can be shown to have developed civilian settlements in the way that frontier forts in Numidia did (Salama 1976). At Aïn Grimidi a village lies around the fort, complete with traces of an enclosure wall with bastions (Gsell 1911, 24.155, plus *additions et corrections*, p 15). Aïn Touta, likewise, has produced traces of a settlement with an irregular enclosure wall (Gsell 1911, 24.119; Cagnat 1913, 663, n3). At Usinaza (Saneg) Gsell and Cagnat recorded a settlement some 300 by 200 m in extent surrounded by a roughly rectangular, stone-faced wall (Gsell 1911, 24.51; Cagnat 1913, 663). At Columnata (Sidi Hosni) traces of a village cover an area *c* 400 × 150 m in size, or nearly 16 acres, although there is no record of defences (Gsell 1911, 22.127).

A plan and record of 1843 shows Tiaret once to have possessed a complex system of defence. This consisted of an innermost wall, *en blocage*, which could have been the original fort, and two outer walls built of large, iron-clamped stones, with closely-spaced bastions (Gsell 1911, 33.14; Cagnat 1913, 660–1, with plan). Gsell thought both of these probably *basse époque*, perhaps constructed by a post-Roman chieftain.

In the far western sector of the frontier, a settlement

of some size is recorded at Lucu (Timiziouine) with some 300 m of wall around part of it (Gsell 1911, 32–46). Altava (Hadjar Roum) we know to have been a *civitas* by AD 220, if not earlier. A 12 ha, 30 acre, enclosure measuring 370×317 m (or, in another source, a little smaller at 380×270 m) is recorded, surrounded by an enclosure wall (Gsell 1911, 31.68; *AE* 1935, 86). A further inscription documents new walls built in 319. Finally, at Numerus Syrorum (Lalla Maghnia), the furthest west fort of Caesariensis (Gsell 1911, 41.1), a rectangular enclosure was recorded in 1887 as being 400×250 m in size, that is something over 11 ha, 26½ acres, and enclosed by a stone wall with square towers at each corner, a gate in the middle of each side flanked by a pair of square towers, and a ditch—one of the very few cases where a ditch has been noted.[19]

At many of these sites, and others such as Aquae Sirensis, Pomaria, and Albulae, evidence, often inscriptions, shows that settlements were occupied until well into the 6th or even 7th century (Gsell 1911, 32.18, 31.56, 9).

The Vandals and after

Late Roman Africa continued into the early 5th century. In AD 429, 80,000 Vandals crossed the Straits: in 430 Hippo fell to a siege which claimed the life of Augustine and in 439 Carthage followed. In 455 the Vandals entered Tripolitania, having defeated all efforts of the Eastern Empire to repulse them.

As an act of policy the Vandals demolished the defences of the cities they took so that they could not be bases for revolt, Carthage and a few places excepted (Procopius, *Bell Vand*, 1.58; *de Aedificiis*, 6.5). This demolition can still be seen at Tipasa. This, then, is the point of time to end a study of town defences in Roman Africa, for what followed, a century later, was the Byzantine conquest, and this has been extensively studied by Diehl (1896) and, more recently, Pringle (1981).

Notes

1 Various ancient descriptions occur, eg Appian, *Punic Wars*, 95; Strabo, 17.3.16; see also Duval 1950; Cintas 1976, 127–31.

2 General and Sabratha: Haynes 1965, 107; di Vita 1968. Leptis: Haynes 1965, 29, 71; Carter 1965; Goodchild & Ward-Perkins 1953. Tripoli: Haynes 1965, 101.

3 Siga later became a Roman *municipium*: Vuillemot 1971, esp 42, 45.

4 Other examples from further afield include: Kalaa (Gsell 1911, 12.102), called *ville romaine*; Gsell's photograph (1901, pl XIII) shows a rocky escarpment with a rough stone wall of huge proportions, approached by a path, while his text talks of two ramparts to the east and one to the west; Souama de Mecherasfa (Gsell 1911, 33.3); Sidi-Medjahed (*Libyca*, 7 (1959), 27–41); Baradez cites Krobzet and the '*camp d'auxiliaires*' (1949, pl 251, 131). Dates are often uncertain.

5 Grenier 1931, 357–61 for colonies in Narbonensis and their walls; Février 1964; 1969.

6 Rusguniae, Gsell 1911, 5.36; Saldae, Gsell 1911, 7.12; Igilgili, Gsell 1911, 7.77; Tubusuptu, Gsell 1911, 7.27; Cartenna, Gsell 1911, 12.20; Gunugu, Gsell 1911, 4.3 and Cagnat 1913, 650.

7 Gsell 1911, 13.28; Ptolemy puts a colony at Aquae, presumably the same site and Salama dates it 33–27 BC.

8 Pringle 1981, 157 notes that square towers are much more common than circular towers in late Roman and Byzantine defences. A comparison of published plans shows that while circular towers are not an exclusively early feature they occur even less frequently in late contexts than Pringle's discussion implies.

9 Thubursicu (Gsell 1911, 18.297), which was described by Tacitus as an *oppidum* (*Ann*, 4.24), grew to a town of *c* 150 acres.

10 Pliny (*Nat Hist*, 5.22) calls it *Colonia Cirta Sitianorum cognomine*; Février 1976, 224–5; Gascou (1972, 142–3) says the first record is AD 103–5.

11 Gsell 1911, 17.126; a photograph is published in Mercier 1903, opposite p 637.

12 These *castella* included Elephantum, Mastarensis, Tidditanorum, Phuensium, Arsacalitanorum, Subzuaritani, and Saddaritanorum (Gsell 1911, 17.93, 94, 89, 102, 271, 276), most of which later became *civitates*.

13 Carcopino (1943, 214–28) expounds the idea of a great uprising, Baradez (1954) brings in the whole of Caesariensis and the Aurès, and Leglay (1959) gives probably the fullest collection of vexillations supposedly rushed to the scene. See Rebuffat 1974, 502 n 2, for other references.

14 Goodchild in Reynolds 1976, 46–58. Fentress (1979, 103) shows Ausum as similar, but is incorrect. The north, east, and west gates have towers with slightly rounded faces, those at the south gate are square.

15 Sherwin-White 1944, 9: 'But Thamugadi became a very large place for the ancient world.' This is hardly borne out by its size; see Figs 1 and 2.

16 Gsell 1901, ch V; '*Pomerium*' and 'Triumphal arch' in *The Oxford Classical Dictionary* (2 edn, 1970).

17 Bu Ngem, Reynolds & Ward-Perkins 1952, 921–2; Gheriat, *Libya Antiqua*, suppl **2**, 94; Ghadames, Reynolds & Ward-Perkins 1952, 910–12 and the 4th century tomb cores.

18 In spite of Matthews (1976, 156), Numidia was downgraded as a command (Birley 1950) and the Mauretanias, unlike Raetia or Noricum, never received a legionary garrison; hardly an indication that security was a major problem.

19 It has been suggested that this might be a fort, but at 26 acres it is much more likely to have been a town.

References

AE *L'Année Épigraphique*, Paris

Allais, Y, 1971 Le quartier occidental de Djemila (Cuicul), *Antiquités Africaines*, **5**, 95–119

Ballu, A, 1903 *Les ruines de Timgad*, Paris

Baradez, J, 1949 *Fossatum Africae*, Paris

———, 1952 *Tipasa*, Algiers

———, 1954 Les nouvelles fouilles de Tipasa... *Libyca*, **2**, 89–149

———, 1966a L'enceinte de Tipasa et ses portes, in Chevallier 1966, 1133–52

———, 1966b Deux amphithéâtres inédit du '*limes*' de Numidie, in *Mélanges offerts à Jérôme Carcopino*, 55–69, Paris

Berthier, A, 1951 *Tiddis*, Algiers

Birley, E, 1950 The governors of Numidia AD 193–268, *J Roman Stud*, **40**, 60–8

Boube, J, 1967 Documents d'architecture mauretanienne au Maroc, *Bull d'Archéol Marocaine*, **7**, 263–367

Brogan, O, 1976 in Stillwell *et al* 1976

Broughton, T R S, 1929 *The Romanization of Africa Proconsularis*, Baltimore

CAH *Cambridge Ancient History*

CIL *Corpus Inscriptionum Latinarum*, **8**, Berlin

Cagnat, R, 1913 *L'armée romaine d'Afrique*, Paris

Carcopino, J, 1925 Le *limes* de Numidie et sa garde Syrienne, *Syria*, **6**, 30–149

———, 1943 *Le Maroc antique*, Paris

Carter, J H, 1965 Western Phoenicians at Lepcis Magna, *Amer J Archaeol*, **69**, 123–32

Chevallier, R (ed), 1966 *Mélanges d'archéologie et d'histoire offerts à André Piganiol*, Paris

Cintas, P, 1976 *Manuel d'archéologie Punique*, **2**, Paris

Courtois, C, 1951 *Timgad*, Algiers

Daniels, C M, 1968 Garamantian excavations: Zinchecra 1965–67, *Libya Antiqua*, **5**, 113–94

———, 1972/3 The Garamantes of Fezzan–an interim report of research, 1965–73, *Soc Libyan Stud Fourth Annu Rep*, 35–40

Diehl, C, 1896 *L'Afrique Byzantine*, Paris

di Vita, A, 1968 Art punique de Tripolitaine, *Mélanges*, Ecole Français de Rome, **80**, 11–83

Duval, P M, 1946 *Cherchell et Tipasa*, Paris

Duval, R, 1950 Enceinte extérieure de la Carthage punique, *Comptes Rendus*, Académie des Inscriptions et Belles Lettres, 53–9

Ennabli, A, 1976 in Stillwell *et al* 1976

Euzennat, M, 1957 Le temple C de Volubilis et les origines de la cité, *Bull d'Archéol Marocaine*, **2**, 41–64

———, 1976 in Stillwell *et al* 1976

Fentress, E W B, 1979 *Numidia and the Roman army*, Brit Archaeol Rep

Février, P-A, 1964 Notes sur le developpement urbain en Afrique du Nord, *Cahiers Archéol*, **14**, 1–47

———, 1969 Enceinte et colonie, *Riv di Studi Liguri*, **35**, 277–86

———, 1976 in Stillwell *et al* 1976

———, 1978 *Djemila*, 2 edn, Algiers

Février, P-A, *et al*, 1970 *Fouilles de Sétif, 1959–66*, 70–86, Paris

Fitz, J, 1976 *Limes*, Akten des XI Internationalen Limescongressus, Budapest

Gascou, J, 1972 Lepti Minus, colonie de Trajan?, *Antiquités Africaines*, **6**, 137–44

Goodchild, R G, 1952 Euesperides–a devastated city site, *Antiquity*, **26,** 208–12
———, 1962 *Benghazi—the story of a city*, 2 edn, Benghazi
———, 1963 *Cyrene and Apollonia*, 2 edn, Libya
Goodchild, R G, & Ward-Perkins, J B, 1953 The Roman and Byzantine defences of Lepcis Magna, *Pap British School Rome*, **21,** 42–73
Grenier, A, 1931 *Manuel d'archéologie Gallo-Romaine*, **1,** Paris
Gsell, S, 1901 *Les monuments antiques de l'Algérie*, **1,** Paris
———, 1911 *Atlas archéologique de l'Algérie, feuille* number followed by site number
———, 1927 *Histoire ancienne de l'Afrique du nord*, **5,** Paris
Haynes, D E L, 1965 *An archaeological and historical guide to the pre-Islamic antiquities of Tripolitania*, Tripoli
Haywood, R M, 1938 Roman Africa, in *Economic survey of ancient Rome* (ed Tenney Frank), **4,** 1–119, Baltimore
Janon, M, 1973 Recherches à Lambèse . . ., *Antiquités Africaines*, **7,** 193–254
Jodin, A, 1966 L'enceinte hellénistique de Volubilis, *Bulletin Archéologique du Comité des Travaux Historiques, 1965–6*, 199–221
Jones, G D B, & Barker, G, 1979/80 Libyan Valleys Survey, *Soc Libyan Stud Eleventh Annu Rep*, 11–36
Jones, G D B, & Little, J H, 1971 Hadrianopolis, *Libya Antiqua*, **8,** 53–67
Kraeling, C, 1962 *Ptolemais, city of the Libyan Pentapolis*, Chicago
Lassus, J, 1966 Une operation immobilière à Timgad, in Chevallier 1966, 1221–31
———, 1969 *Visite à Timgad*, Algiers
———, 1976 in Stillwell *et al* 1976
Leglay, M, 1959 Une nouvelle inscription relative a l'insurrection mauretanienne de 145–147, *Libyca*, **7,** pt 2, 217–20
———, 1963 Les Flavians et l'Afrique, *Mélanges*, Ecole Français de Rome, **80,** 201–46
Le Glay, M, 1968 A la recherche d'Icosium, *Antiquités Africaines*, **2,** 7–52
Leveau, P, 1972 Aile II des Thraces, Mazices et Praefecti Gentes, *Antiquités Africaines*, **7,** 186–9
Lézine, A, 1966 Utique, notes de topographie, in Chevallier 1966, 1241–55
———, 1971 Utique, note d'archéologie Punique, *Antiquités Africaines*, **5,** 87–94
Lloyd, J A (ed), nd *Excavations at Sidi Khrebish, Benghazi*, **1,** *Libya Antiqua*, suppl 5
Luquet, A, 1972 *Volubilis*, Tangier
Marec, E, 1954 *Hippone la Royale*, Algiers
Marion, J, 1957 Les ruines anciennes de la region d'Oudja, *Bull d'Archéol Marocaine*, **2,** 117–73
Matthews, J, 1976 Mauretania in Ammianus and the Notitia, in *Aspects of the Notitia Dignitatum* (ed R Goodburn & P Bartholomew), Brit Archaeol Rep, 157–88
Mercier, E, 1903 *Histoire de Constantine*
Moscatti, S, 1973 *The world of the Phoenicians*
North Africa entry in *Encyclopedia of World Art*, **1** (1959), 66–130
Picard, G C, nd *Castellum Dimmidi*, Paris
Poinssot, C, 1958 *Les ruines de Dougga*, Tunis
Pringle, D, 1981 *The defence of Byzantine Africa from Justinian to the Arab conquest*, Brit Archaeol Rep
Rebuffat, R, 1969/70 Bu Ngem, 1968, *Libya Antiqua*, **6/7,** 21–2
———, 1974 Enceintes urbaines et insécurité en Mauretanie Tingitane, *Mélanges*, Ecole Français de Rome, **86,** 501–22
Reynolds, J M (ed), 1976 *Libyan studies: select papers of the late R G Goodchild*
Reynolds, J M, & Ward-Perkins, J B, 1952 *The inscriptions of Roman Tripolitania*
Romanelli, P, 1970 *Topographia e archeologia dell' Africa Romana*, Enciclopedia Classica, Turin
Salama, P, 1953/5 Nouveaux témoignages de l'oeuvre des Sévères dans la Mauretanie Césarienne, *Libyca*, **1,** 231–61; **3,** 329–67
———, 1966 Occupation de la Mauretanie Césarienne occidental sous le Bas-Empire romaine, in Chevallier 1966, 1291–1311
———, 1976 Les displacements successifs du *limes* en Mauretanie Césarienne, in Fitz 1976, 577–96
Seston, W, 1928 Le secteur de Rapidum . . ., *Mélanges*, Ecole Français de Rome, **45,** 150–83
Sherwin-White, A N, 1944 Geographical factors in Roman Algeria, *J Roman Stud*, **34,** 1–9
Stillwell, R, MacDonald, W L, & McAllister, N H, 1976 *The Princeton encyclopedia of classical sites*
Syme, R, 1951 Tacfarinas, the Musulamii and Thubursicu, in *Studies in Roman economic and social history in honor of Allan Chester Johnson* (ed P R Coleman-Norton), 113–30, Princeton
Trousset, P, 1976 Le camp de Gemellae sur le limes de Numidie d'après les fouilles de Colonel Baradez, in Fitz 1976, 559–76
Vuillemot, G, 1971 Siga et son port fluvial, *Antiquités Africaines*, **5,** 39–86
Ward-Perkins, J B, 1976 in Stillwell *et al* 1976
White, D, 1976a in Stillwell *et al* 1976
———, 1976b *Apollonia, the port of Cyrene, Libya Antiqua*, suppl **4**

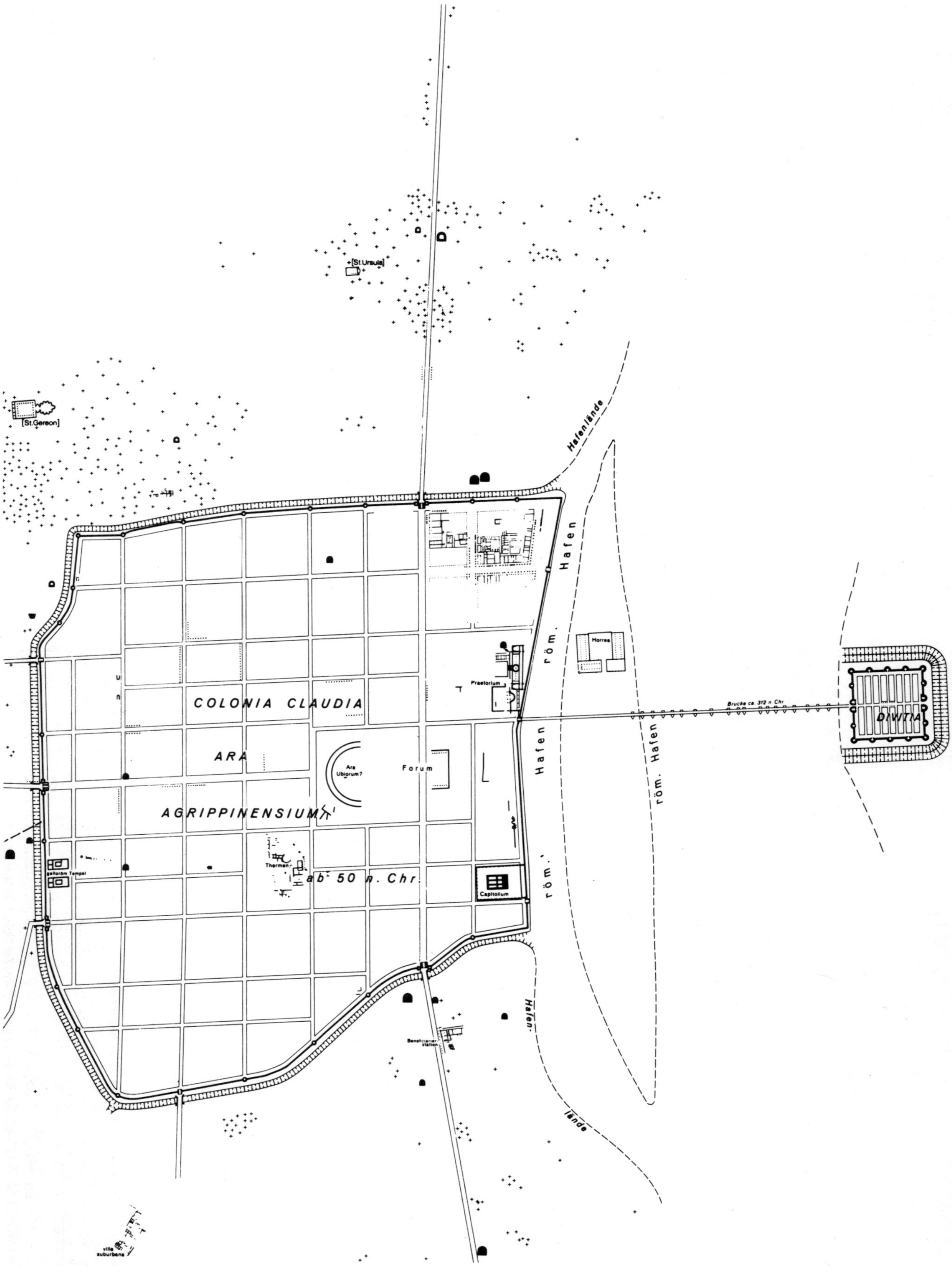

Fig 6 Plan of Colonia Claudia Ara Agrippinensium (CCAA), featuring the city wall and ditch

The Roman defences of Cologne—Colonia Claudia Ara Agrippinensium

Hansgerd Hellenkemper

Ara Ubiorum

Topography and the founding of the oppidum

The Roman settlement of Cologne was situated on the west bank of the Rhine in the shadow of the mountains of Rhenish Schist which lie to the south. The approximately 1 km square site on a raised gravel plateau, 4–5 m above the surrounding area, was chosen for strategic reasons. The gravel plateau, with a tributary of the river flowing to the south, was also bordered by a narrow strip of flood plain which widened into a broad river bed further downstream. In the last decades of the 1st century BC, the Ubii were forced to settle on the plateau in an *oppidum* planned by the Romans. The settlement was known as Ara Ubiorum (Fig 6). The *oppidum* had a street grid laid out in the Augustan period, with timber-framed houses and public buildings of stone, which was brought from outside.

The situation and size of the camp *apud Aram Ubiorum* of the two legions (*Legio* I and *Legio* XX) whose presence here in AD 14 is cited by Tacitus (*Ann*, I. 31ff) is disputed (Filtzinger 1962/63; Precht 1971, 53ff; Bracker 1974, 113–26; Hellenkemper 1972/3, 59ff; 1975, 786–8).

The earth and timber rampart

Only on a few occasions have the defences of the *oppidum* (or possibly those of the legions?) been traced: by the north gate of the *colonia* (see below) stratigraphy was found which suggested the presence of an earth and timber rampart (Doppelfeld 1950, 12–21). To the west, at the gate of *decumanus maximus*, there was similar stratigraphy. The later stone wall was evidently in the same position as the earth and timber rampart in both instances (Doppelfeld 1950, 20–2; critically reviewed by von Petrikovits 1952).

These defences were not located to the south but a foundation of large tufa blocks by the southern harbour tower indicates that the rampart can once again be expected close to or under the later stone wall.

An accompanying ditch can be presumed on an identical alignment to its counterpart associated with the stone wall. To the east the Rhine is a natural boundary and there is no early evidence of defences in this area. A short row of wattles (dated to *c* AD 24 by dendrochronology: Hollstein 1980, 73) might represent a waterfront.

The harbour tower

The oldest known stone building of the Roman Rhineland was found in 1966, in the south-east corner of the *oppidum*, and conserved (Bracker 1974, 126–32; Hellenkemper 1975, 788–9) (Fig 7). The bottom of the foundation of the squarish stone tower lay some 6 m below the later *colonia* horizon. The foundation was prepared by

Fig 7 The southern harbour tower, c *AD 4*

Fig 8 North-west corner tower, c mid 1st century AD, as recorded in 1836

Fig 9 North-west corner tower, as recorded in 1980 (with battlements added before 1906)

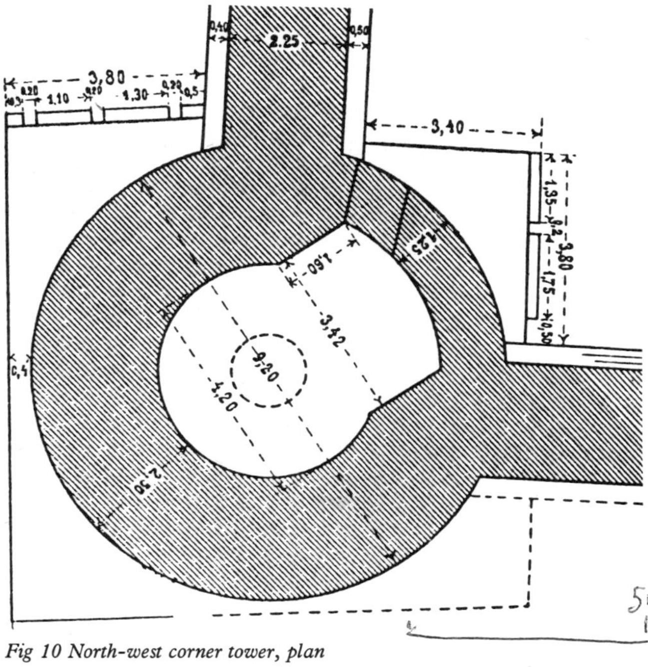

Fig 10 North-west corner tower, plan

driving oak timbers about 2 m long into the natural gravel. These piles were 0·10–0·12 m in diameter and spaced at about 0·40 m intervals, forming an almost regular grid with which to consolidate the ground. A raft of *opus caementicium* with sandstone chips, measuring some 10·90×10·60 m, and 0·22 m in depth, covered the wooden grid. On this base the foundation of tufa blocks was laid, 9·70×9·38 m in area, which consisted of three courses of varying height. Up to nine courses of tufa blocks of the upstanding wall survive and it is clear that the tower was already in a poor state of repair in the mid 1st century. Recent dendrochronological analysis of the piles suggests a construction date of *c* AD 4 (Hollstein 1980, 72–3).

A 19th century archaeological survey indicates that there was a similar tower at the north-east corner; in common with the south-east tower where the stone blocks were formed into a T-shaped foundation, facing west, it seems that a similar technique of construction was employed at the north-east corner. Since there were no signs of a wall extending north of the south-east corner tower, a riverside wall along the Rhine can be discounted.

Colonia Claudia Ara Agrippinensium

The establishing of the colonia

In AD 50, Claudius raised the *oppidum* Ara Ubiorum to the rank of a *colonia* and the settlement was given the official title Colonia Claudia Ara Agrippinensium (CCAA) (see Fig 16). This promotion seems to have provided the impetus for a programme of civic building in which Claudian/Neronian and Flavian phases can be recognized. Manufacturing industries which involved a fire risk were restricted to extramural areas. It is probable that soon after the status of *colonia* was granted, a defensive stone wall was built on the alignment of the earth and timber rampart (Fig 6).

The defensive circuit

The city wall was some 3911·8 m long (N: 948·9 m, W: 1177·9 m, S: 903·3 m, E: 881·7 m) and enclosed an area of 96·8 ha (Schultze & Steuernagel 1895, 8, pl 1) (Fig 6). It followed exactly, as far as is known, the edge of the plateau. Only the eastern side lay further out in the Rhine water meadows, close to the tributary. There was a second wall built in several sections, not for military reasons, but as a support for the 4–5 m high slope of the plateau terrace.

The city wall had nineteen towers (distances between centre points varied from 77 m to 158 m), positioned according to topographical factors (Schultze & Steuernagel 1895, plan) (Figs 8, 9, 10, 11, 12), and nine gates of different designs. The whole length of the wall facing the Rhine was without towers, as were its north and south corners (see above). There were three gates on the west, two on the south, three on the east, and one on the north, and their varying dimensions were obviously closely related to their significance in terms of long-distance and local traffic. On the external side of the wall, on the three landward sides, was a ditch.

Since the 18th century, 70% of the entire circuit has been recorded, and at the moment a comprehensive report is being produced (Süssenbach 1980).

The gates

Nine gates are known (Schultze & Steuernagel 1895, 34ff; 1916, 1ff; Doppelfeld 1950, 9ff; 1961, 35ff), and this is thought to be the original number. Although only the foundations of most gates have survived, their relative importance can be seen in terms of their varying sizes. To the south, west, and north, on the axis of the *cardo* and *decumanus maximus*, lay the main gates, each with three arches. They provided access to the major roads, together with a gate of two arches. The recovery of the plan of a gate on the north has allowed a reconstruction to be made (Figs 13, 14, 15, 16, 17). This gateway of three arches must originally have been two storeys high because of the space needed to lift the portcullis. The conjectural plan of the upper floor can only be based on that of the ground plan. The facade at ground level has been reconstructed, featuring four Corinthian pilasters. Details from sculptured stonework found in the vicinity were incorporated by Schultze into his reconstruction. However, Kähler (1942, 95ff) does not accept this reconstruction, because of his knowledge of gates in Italy (although Mylius (1955, 55ff) is partly convinced). Nevertheless, a date in the 1st century AD (as for the city wall) seems to be safe. On the basis of examples in Italy, Kähler suggests that the reconstruction should feature an upper storey with windows and eight half-columns or pilasters, above the three pilasters of the ground floor.

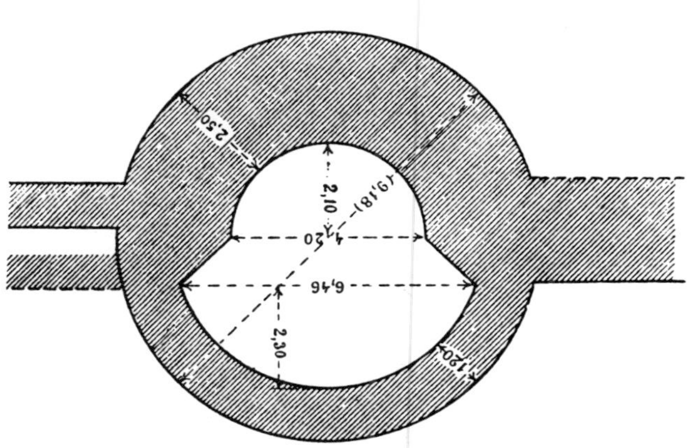

Fig 11 Common type of tower, north city wall

Fig 12 Tower and a stretch of the city wall

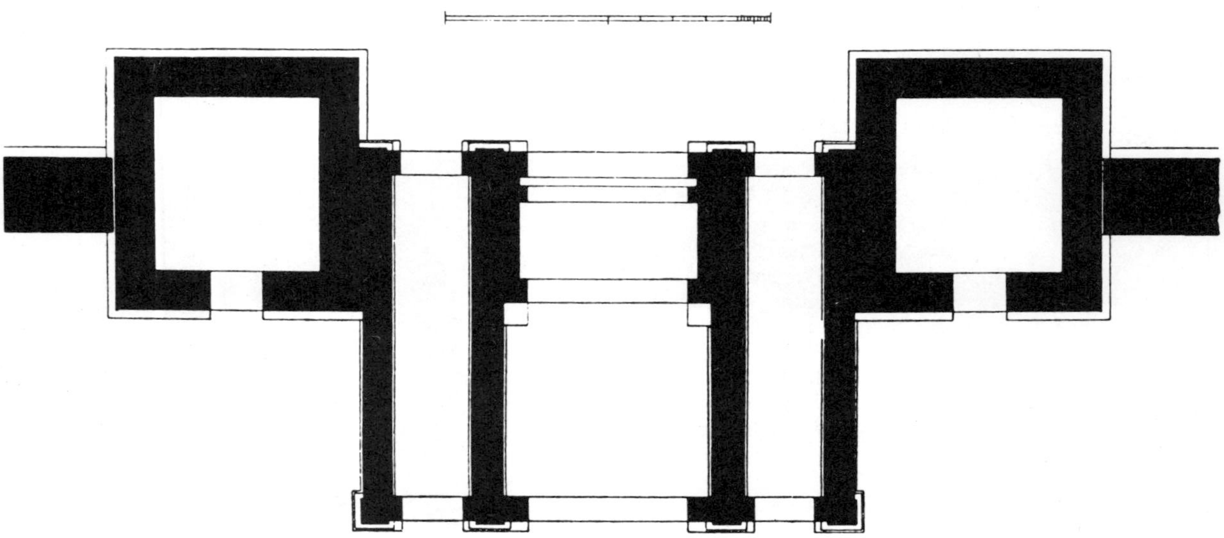

Fig 13 North gate, plan at ground floor level (Mylius 1955)

Fig 14 North gate, intramural face (model by H Schultze)

Fig 15 North gate, extramural face (reconstruction—Mylius 1955)

Fig 16 North gate, middle arch with CCAA inscription above eradicated inscription of c *AD 254–9*

The ditch

The ditch has been recorded on the three landward sides of the *colonia*. No traces of a double or treble ditch system have been found and are unlikely to have existed since suburban buildings were situated only 26 m outside the city wall. The berm separating the ditch from the wall was between 3 and 4 m, and the ditch was some 13 m wide and up to 3·30 m in depth; the bottom was 0·80 m wide. It is not known for how long the ditch was maintained and open, but there are indications of two periods of use in some places.

Materials and construction techniques

The city wall, including towers and gates, was built of stone bonded with lime mortar. The proportions of types of stone used have been roughly estimated by observation: about 80% sandstone with schist for wall cores and facings, 5% basalt, and 3% trachyte. Red sandstone and limestone were used for the gates and also, presumably, for the parapet wall of the walkway on top of the city wall. Tufa does not seem to have been used, perhaps because it is rather soft.

Because there is no stone available in the area, it had to be entirely imported from quite a distance via waterways. The sandstone (containing schist) may have been brought from the middle Rhine, although the actual quarries are not known. Basalt and trachyte came from quarries close to Drachenfels, the red sandstone from around Trier/Mosel, and the limestone from quarries in Norroy/Lorraine. These quarries were discovered by Roman soldiers and may also have been exploited by them. The lime may originate from the Erft area where *Legio I Minervia* had a factory. The sand came from pits in front of the wall—one sandpit of 1st century AD date has been found on the north side. For the foundations, trenches with sharply-angled sides were dug up to 0·50 m into the fine sand layers of the subsoil. Shoring timbers and planks were sometimes placed in the bottom and left there. The underlying gravel was presumably not reached by the foundation trenches and there is no information of the extent to which the foundations were supported by the wooden piles, although this method of construction was used for the north gate.

A continuous grid of wooden piles for the eastern side of the city wall, facing the flood plain, is probable but could not be proved in recent investigations.

The base course of the foundations was often composed of upright drystone blocks of sandstone, basalt, and trachyte rubble. Above this, and between two irregular faces of rubblework, *opus caementicium* was poured forming compacted layers. The facing courses vary from 0·13 to 0·20 m in height. The first row of scaffolding holes (0·10 m high, 0·15 m wide) appears at a height of 2·30 m on the western side of the city wall. No scaffolding holes have been observed in the upper part of the wall and obviously they were carefully filled in after the scaffolding was dismantled. The internal and external wall faces were visible from the level of the footings and all the visible city wall and tower faces were uniformly constructed of sandstone, average size 0·10 m in height by 0·25 m wide. Traces of horizontal settling in the core of the wall indicate that three rows of the facing work were constructed at a time and then the core of small rubble-work in *opus caementicium* was built up.

Despite the care taken with the foundation it seems that some settlement cracks may have occurred during construction.

All the investigations have indicated that the city wall and the towers were constructed at the same time, both in foundation and superstructure. Only the foundations of a few of the gates were constructed separately and it is clear that these were laid before the foundations of the city wall.

Discussion and conclusions

Historical perspective

No other Roman monument of the *colonia* has been so extensively investigated and discussed in modern times as the city wall. Until the 11th century AD, this wall provided the only secure defence and for that reason it was not systematically robbed for building material as were other important buildings (eg the praetorium, forum, temples, baths, etc). The construction of the city wall was the most extensive undertaking in the *colonia* in the Roman period.

Despite careful investigations since the second half of the 19th century and several detailed discussions (Schultze & Steuernagel 1895; Schultze 1909; Kraus 1926; Fremersdorf 1934; Doppelfeld 1950), there are many problems remaining. One reason for this is the dilemma of modern urban archaeology, where more opportunities exist for rescue—rather than research—excavations. No section of the Roman wall has been comprehensively investigated: all summaries are drawn from a multitude of the different observations during the last century.

Dating

The lack of opportunity for research excavations has meant that the city wall can only be dated approximately. Nevertheless, a number of facts have indicated a likely period (Schultze & Steuernagel 1916, 1ff; Fremersdorf 1934, 64ff; Kähler 1942, 95ff); eg the plan and face of the

Fig 17 North gate, side entrance, reconstructed near the findspot

north gate compare with early Roman examples on stylistic grounds. The same may be said of the towers. Investigation of the city wall itself and the fill of the construction trenches has not so far made further definition possible. However, the technique of masonry construction used can be compared with parts of the praetorium which have been dated (Precht 1973). In conclusion, a date for the construction of the defensive circuit in the 1st century AD seems probable.

Reconstruction

Discussion of the date of the defensive circuit has influenced the question of reconstruction. There is no definite proof of supplements or repairs and the inscription in the middle arch of the north gate—[[VALERIANA GALLIENA]]—is presumably an addition of *c* AD 254–9, which was perhaps eradicated in the reign of the usurper Postumus (Fig 16). Any trace of rebuilding may no longer exist, since few upstanding sections of the upper part of the wall have survived. There is, however, a documentary reference to a rebuilding of the wall after a raid by the Normans in AD 881.

Origins

The plans of the towers and gates clearly indicate their architectural design. The city wall of the *colonia* was of uniform construction based on a clear, careful design. The designs did not originate in the German provinces since there were no examples of such fortifications in this area in the middle of the 1st century AD. Kähler, Ward-Perkins, and Gabelmann have emphasized the strong influence of the towns of upper Italy on the diffusion of Roman art and architecture during the late republic and early empire. It appears that the design of the city wall of the *colonia* was influenced in this way.

The ornaments which seem to have featured haphazardly on the city wall and towers—and are visible at several places—can obviously be traced back to reticulated walls in Italy (see Fig 8). There are several parallels in Gaul, though not securely dated, and their origin and diffusion have to be investigated more closely.

References

Bracker, J, 1974 Neue Entdeckungen zur Topographie und frühen Geschichte des römischen Köln, *Jahrb des Kölnischen Geschichtsvereins*, **45**, 111–74

Doppelfeld, O, 1950 Die römische Stadtmauer von Köln, in *Die Kunstdenkmäler im Landesteil Nordrhein*, Beiheft 2, *Kölner Untersuchungen* (ed W Zimmermann), 3–40

———, 1961 Das neunte Tor von Köln, in Miscellanea archaeologica in hororem J Breuer, *Archaeol Belgica*, **61**, 35–44

Filtzinger, Ph, 1962/63 Zur Lokalisierung der Zweilegionsfestung 'apud Aram Ubiorum', *Kölner Jahrb für Vor- und Frühgeschichte*, **6**, 23–57

Fremersdorf, F, 1934 Das Alter der römischen Stadtmauer Kölns, *Bonner Jahrb*, **139**, 64–79

Hellenkemper, H, 1972/73 Oppidum und Legionslager in Köln. Überlegungen zur frührömischen Topographie, *Kölner Jahrb für Vor- und Frühgeschichte*, **13**, 59–64

———, 1975 Architektur als Beitrag zur Geschichte der Colonia Claudia Ara Agrippinensium, in *Aufstieg und Niedergang der römischen Welt II Principat*, **4**, 783–824, Berlin

———, 1979 *Deutscher Städteatlas*, 2,6 (ed H Stoob), Münster

Hollstein, E, 1980 Mitteleuropäische Eichenchronologie. Trierer dendrochronologische Forschungen zur Archäologie und Kunstgeschichte, *Trierer Grabungen und Forschungen*, **11**, 72–3, Mainz

Kähler, H, 1942 Die römischen Torburgen der frühen Kaiserzeit, *Jahrb Deutschen Archäologischen Institutes*, **57**, 1–104

Klinkenberg, J, 1906 Das römische Köln, in *Die Kunstdenkmäler der Stadt Köln*, **1**,2 (repr 1980), 131–393

Kraus, O, 1926 Die römische Rheinbrücke zu Köln und die Stadtmauer der Südseite, *Bonner Jahrb*, **130**, 232–53

La Baume, P, 1972 Das römische Köln, *ibid*, **172**, 271–92

Mertz, M, 1883 Beitrag zur Feststellung der Lage und der jetzigen Beschaffenheit der Römermauer zu Köln, in *Programm der Ober-Realschule zu Köln . . . für das Schuljahr 1882–83*, 3–28

Mylius, H, 1955 Des Nordtor der Colonia Agrippinensis, *Kölner Jahrb für Vor- und Frühgeschichte*, **1**, 9–16

von Petrikovits, H, 1952 review of Kölner Untersuchungen, *Germania*, **30**, 125–31

Precht, G, 1971 Die Ausgrabungen um den Kölner Dom Vorbericht über die Untersuchungen 1969/70, *Kölner Jahrb für Vor- und Frühgeschichte*, **12**, 52–64

———, 1973 Baugeschichtliche Untersuchungen zum römischen Praetorium in Köln, *Rheinische Ausgrabungen*, **14**

Schultze, R, 1909 Die römischen Stadttore, *Bonner Jahrb*, **118**, 280f

Schultze, R, & Steuernagel, C, 1895 Colonia Agrippinensis, *ibid*, **98**, 1–144

———, & ———, 1916 Neue Beiträge zu Colonia Agrippinensis, *ibid*, **123**, 1–14

Süssenbach, U, 1980 Plan der römischen Stadtmauer Kölns, *Führer zu vor- und frühgeschichtlichen Denkmälern*, **37**.2, Köln I.2, 72–5, Mainz

———, 1981 *Die Stadtmauer des römischen Köln* (published after this paper was written)

The town walls and defensive systems of Xanten—Colonia Ulpia Traiana

Gundolf Precht

Introduction

Archaeological investigations of the site of Colonia Ulpia Traiana (CUT) have been carried out since the second half of the 19th century, before which time buildings and other structures were over the centuries systematically plundered for the sake of their stone.

The first phase of the investigations concerned the town wall and its course, as well as areas within the town itself which seemed from earlier observations to have potential. Parts of the public baths, the forum, and the amphitheatre in the south-east corner of the town were excavated (Steiner 1911, 10).

The full extent of the settlement was ascertained through a series of excavations undertaken by the Lower Rhine Historical Society. During the construction of the railway from Moers to Cleves (Kleve), Koenen discovered a large gateway flanked by two towers on the western side of CUT. The main gate on the northern side was partially excavated by Steiner (1903, 182–7).

Between 1934 and 1936 the Rheinisches Landesmuseum investigated the eastern part of the town adjacent to the wall. We are indebted to von Petrikovits and Stoll for major discoveries made under their supervision; a wooden quay running along the eastern side of CUT proved to be of early construction (von Petrikovits 1952, 133, 138), and the first evidence of a harbour temple also came to light (*ibid*, 125). The amphitheatre in the south-east corner of the town, which had been revealed in 1888, was fully exposed (von Petrikovits 1937, 248; Heidenreich 1940, 33–62). The town wall was revealed in these two areas but not examined to any extent.

After the Second World War the Rheinisches Landesmuseum concentrated its efforts on certain internal parts of the town which were to be redeveloped. A large amount of information was gathered about the structures and the plan of the town. However, with the exception of a few rescue digs no new investigations of the wall itself were undertaken. On the southern side a ditch and a stretch of wall were examined, though their interpretation remains unclear. Since the latter was only 0·80 m wide, it would seem unlikely that it was part of a fortified structure, even assuming it to be of Roman origin. Another somewhat wider wall in front of the actual town wall was found during the excavation of the southern gate in 1933 (Stoll 1936, 431).

Systematic investigation of the town wall has been in progress since the establishment of the Archaeological Park in 1974 in the eastern sector of the ancient town, for the purpose of reconstructing a section of the defensive system (Fig 18; Precht & Rupprecht 1976, 338–51). The results have been surprising in certain respects, especially with regard to the pre-colonial settlement in this area. But before discussing the pre-colonial settlement, we should first consider the defensive system of the colony itself. For the sake of convenience I shall discuss in turn (1) the northern wall, (2) the southern wall, (3) the eastern or river-front wall, and (4) the western wall.

It must be emphasized that no part of the defensive circuit has been preserved above ground, so that its construction can only be determined by excavation. Since the course of the town wall was established largely in the 19th century, the most recent investigations have been concerned with assessing the defensive nature and date of the wall.

The northern wall

The northern wall has been sectioned in several places between the north-east corner and the main gateway which was examined in the 19th century (Precht 1976, 68).It survives just beneath the modern ploughsoil and was approximately 1·75 m wide. The bottom of the foundation consisted of a course of 'greywacky' stone set in a packing of a clayey loam, which was some 0·70 m thick in places. Above this was a layer of 'greywacky' chippings in a hard gravelly cement (cf Fig 23). The wall foundation was set into natural and was not supported by piles (see below). The walling had not survived above its foundation level.

In the angle of the eastern and northern walls stood the anticipated corner tower (Fig 19; Precht 1976, 71). It had been completely robbed out so that its ground-plan could only be deduced from the robbing trenches. As in the cases of the south-east and north-west corner towers (Scollar 1965, Tafel 24b; see also Steiner 1903, 185), the latter having been ascertained by means of aerial photography, it was trapezoidal in shape and founded on piles. It straddled the rounded corner of the wall and its external face projected for some 0·60 m. Approximately 120 m to the west another tower was revealed, which had also been seen in the 19th century (Figs 20, 21). Once again this was set upon a hollow foundation approximately 6 m by 6·6 m which projected 0·90 m beyond the wall. A box-drain had been laid upon its transverse axis, but there was no confirmation that this might indicate an opening for a door or gateway.

An extensive area in front of the tower was investigated (Precht 1977, 163) and a ditch located with the remains of a 2·20 m wide cement foundation in the bottom. By analogy, it represents the remains of a substantial sewer which was probably constructed of large tufa blocks laid upon a cemented foundation course. The box-drain beneath the tower would have run into this main sewer. As in the case of the southern wall, the ditch had mostly been obliterated during the course of centuries of ploughing, and indeed even the foundation of the wall did not survive in this area. The details of the foundation of the northern wall, as far as may be assessed, correspond well with those of the southern wall.

The southern wall

Between the corner tower and the next interval tower to the west, the superstructure of the wall had been entirely robbed out but the foundation was of similar construction to that on the northern side (Precht & Rupprecht 1976, 338–51; Precht 1975, 57).

The south-east corner tower had also been entirely robbed out but despite this the trapezoidal shape was clearly definable, being between 7 m to 6·20 m and 5·50 m to 5·80 m in extent. This tower must therefore have had a similar ground-plan to that of the north-east corner tower. It also projected 60 cm beyond the rounded corner of the wall (Figs 19, 22).

The plan of the nearest interval tower to the west resembled that of its opposite number to the north, although

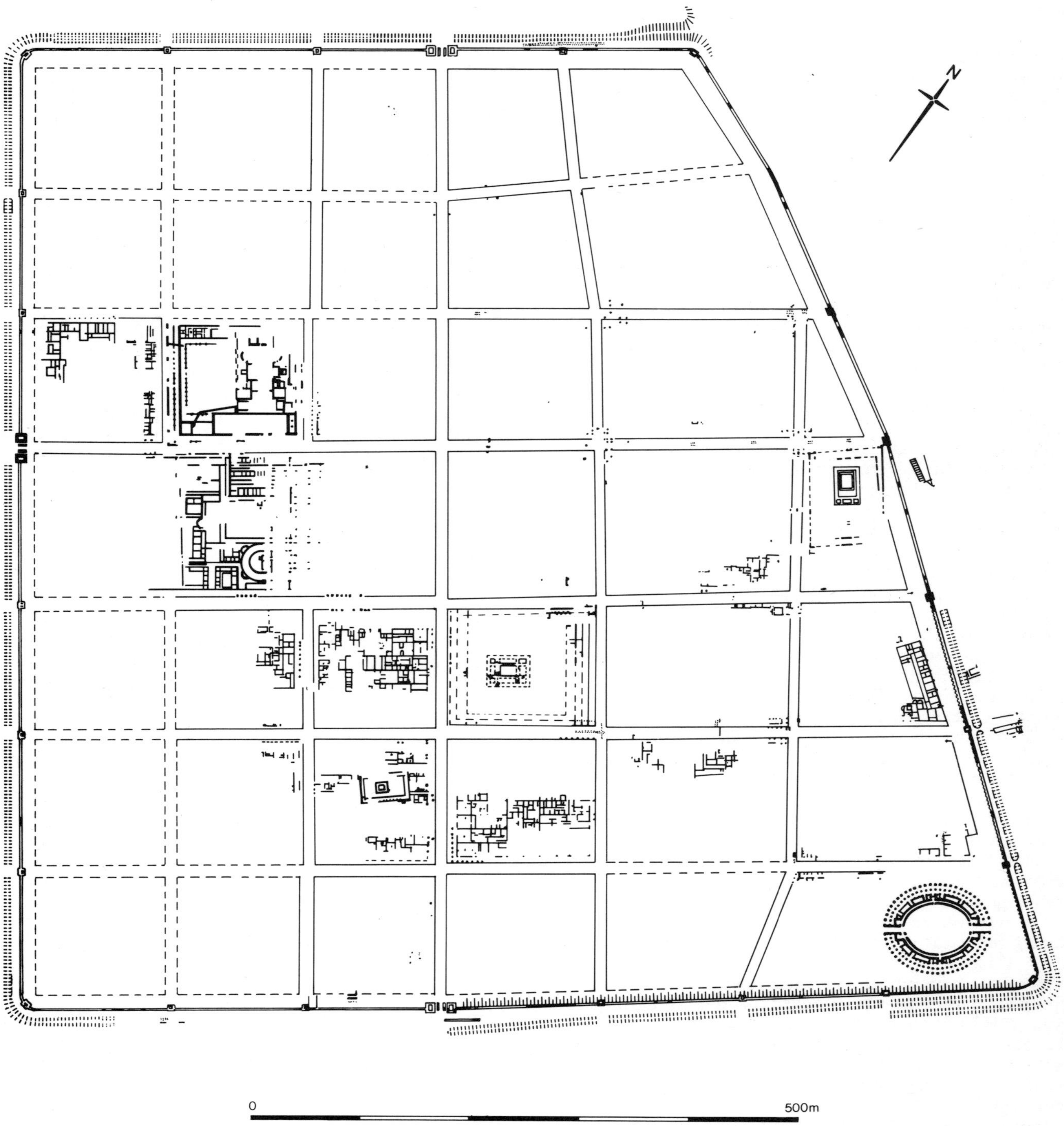

Fig 18 Colonia Ulpia Traiana: plan of the town (1979)

Fig 19 South-east corner tower

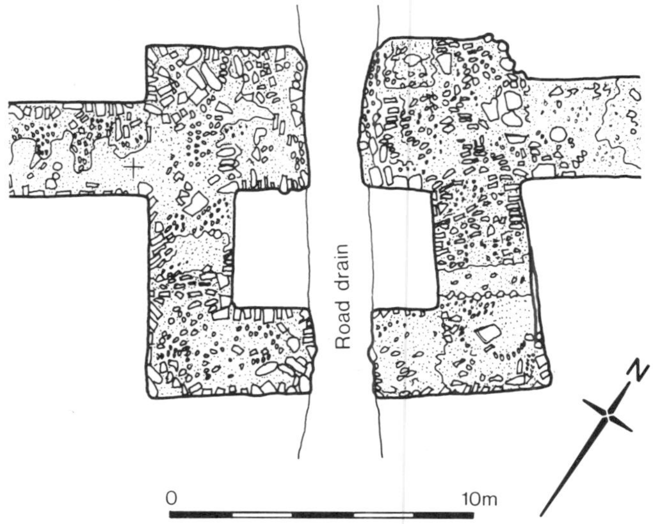

Fig 20 Interval tower in the northern wall

Fig 21 Interval tower in the northern wall with break for drainage channel

Fig 22 Reconstruction of south-east corner tower from exterior of town

Fig 23 Foundation of the town wall, interval tower on south side

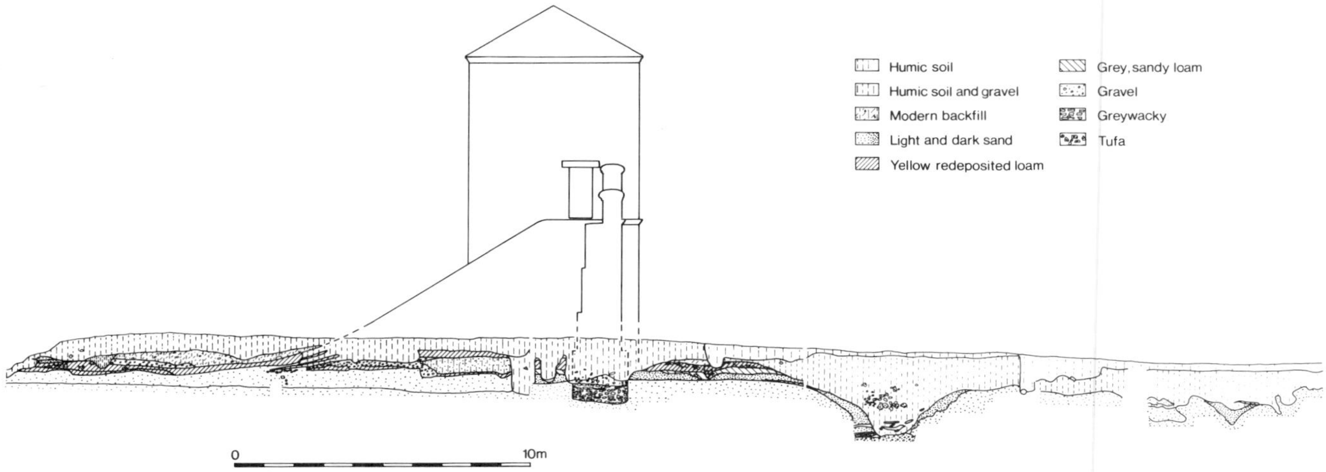

Fig 24 Cross-section through the southern wall and ditches

only part of the foundation was seen (Figs 20, 23). It was roughly 6 m by 5·50 m in size and was thus slightly smaller. On its northern side was found a large amount of fallen slates which, considered together with similar material elsewhere around the towers, indicate the type of roof construction.

A north-south trench against the inside face of the wall showed in section a series of layers which appeared to be the remains of a bank or rampart (Figs 24, 25). To the south of the wall were found large waste deposits of tufa including whole ashlar blocks, probably representing a working area associated with the wall's construction. If so, it can be suggested that the wall's superstructure consisted of a tufa facing against a core, a large block of which was found in the upper layers of the inner ditch.

Outside the wall were two V-shaped ditches approximately 3 m deep and 7 m wide (Precht 1975, 58). Both ditches followed the curve of the south-east corner. The outer ditch stopped approximately 40 m north of the corner tower, whereas the inner ditch, despite its proximity to an ancient tributary of a river, continued northwards although reduced in width (Fig 26) and may have terminated near the quay. A second wall, which had been observed in earlier excavations in front of the southern wall and its western extremity, could not be seen in the south-east corner. In the inner ditch was found a quantity of sculptured fragments of tufa, which were at first taken to be the remains of a moulding on the exterior face of the wall at the height of the parapet (Precht 1975, 58). Recently, in the vicinity of the ditches, a similar tufa moulding came to light; it proved indisputably to be part of a coping-stone with a carved cornice (Fig 27). The slab has a slightly curved profile indicating that it originated from the rounded corner of the wall.

Fig 25 Reconstruction of south-east corner tower from within the town

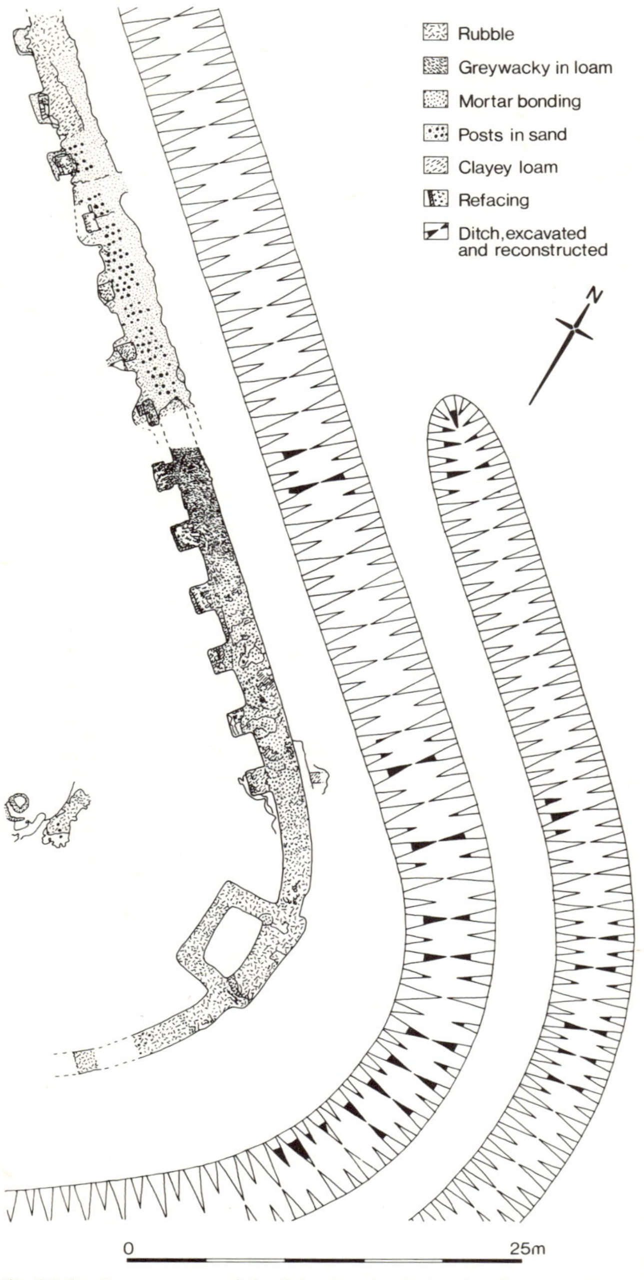

Fig 26 South-east corner of the defensive circuit showing the corner tower and ditches

The eastern wall

In recent years it has become clear that the history of the construction of the eastern wall is more complicated than was at first appreciated.

Surprisingly, the walls and towers were not built together, but must belong to at least two phases (Precht & Rupprecht 1976, 338–51; Precht 1975, 58; 1977, 160).

The southern section of the eastern wall must belong to the same period as the southern and northern walls. Its foundations were some 1·80 m wide but were strengthened internally with buttresses. These were between 1·70 m and 1·80 m broad and between 1·30 m and 1·50 m wide at intervals of approximately 4·15 m (Figs 26, 28). The material and method of construction correspond with the previously described remains, so far as the scanty evidence can be judged. The buttresses which

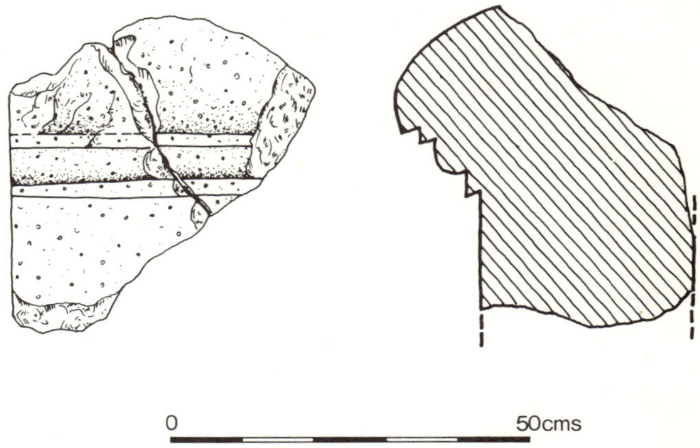

Fig 27 Coping stone, original and cross-section

survived were of differing dimensions and slightly inclined towards the wall, tapering towards the top, and supposedly also acted as supports upon which the substructure of a parapet could have rested (Fig 29). The foundations on this side were laid on a grid of piles, owing to the ancient tributary of the Rhine which was just outside the wall (Fig 26). No rampart existed on this eastern side.

The foundation of the north part of the wall survived far better than elsewhere. It was about 3·50 m wide, consisting mainly of basalt, and as in the northern wall contained two courses: cement poured over a clayey base. The foundation was partly set into the muddy deposits of the river bank, in which the oak piles were well preserved. A row of piles with the bark still adhering was recovered and examined; according to the dendrochronological results the trees were felled in the winter of AD 105 (oral confirmation from Dr Schmidt of the Institut für Dendrochronologie, Cologne University), so for the first time there exists a precise date for the construction of the wall. Three towers may be assigned to this phase of construction, and, as opposed to those resting upon the dry ground of the landward sides of the colony, they had solid foundations which comprised blocks of basalt bonded with mortar, supported by basalt set into clay. The largest tower (9·00 m by 7·25 m) lay on the axis of the *decumanus maximus*, and,

Fig 28 Eastern wall, lowest foundation layer

Fig 29 Reconstruction of eastern wall

although evidence of its superstructure is lacking, it might be part of a gateway (Figs 30, 31). It had previously been suggested that a gate flanked by two towers stood on this site but this still cannot be proved. About 120 m to the south was built the so-called 'Mole Tower', which had a virtually square ground-plan some 7·45 m by 7·50 m (Fig 32). Except for a small corner of the basalt foundation, the masonry had been totally robbed, yet because of an underlying pile grating it can be deduced that it too had a solid foundation.

Information concerning the chronological sequence of construction first came with the investigation of what we today call the 'Small Harbour Gate' (Fig 33). The excavation left little doubt that a larger tower was to have been built upon an oblique oblong-shaped cemented basalt base, and a foundation platform for this was constructed. However, eventually a smaller tower (some 6 m by 6 m) was built, of the type already described on the southern side, and the solid platform was removed in the area of the new foundation. A large sandstone block (some 0·60 m by 0·90 m by 1·50 m) was found above the platform; one corner had a worn surface and scratch-marks indicate that it was a base block of a portal or gateway, through which wheeled vehicles passed.

The same situation is true of the interval tower erected further to the south, ie north-east of the amphitheatre. A smaller hollow foundation of 'greywacky' stone some 6 m by 6 m was laid upon the remains of a basalt platform.

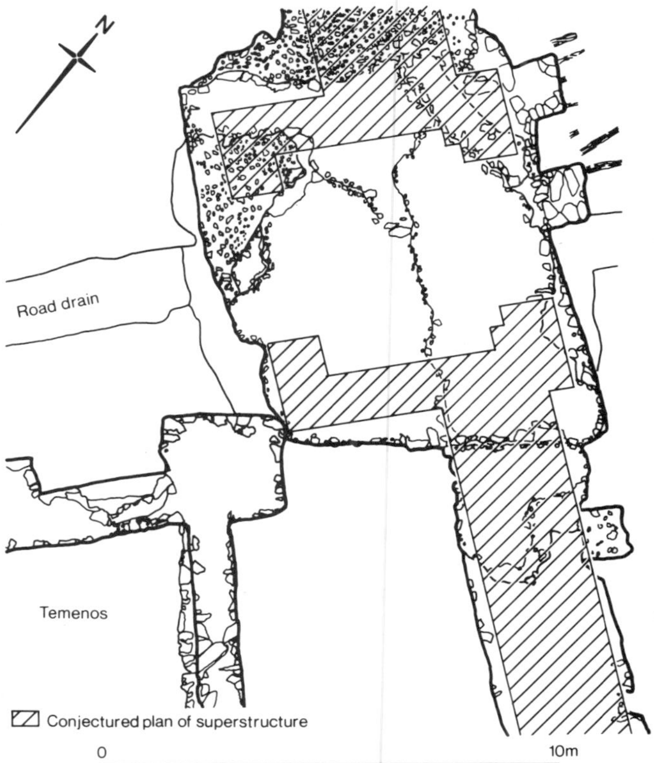

Fig 30 Large Harbour Gate, ground plan

Fig 31 Foundation of Large Harbour Gate, with corner of precinct wall of harbour temple

The western wall

According to earlier observations, the wall on the western side should be of uniform construction, having a foundation of 'greywacke' between 1·78 m and 2 m in width. Apart from one excavation on the site of the main gate in the 19th century, no archaeological record of the towers exists; it can be assumed on existing evidence, however, that they were situated at the end of the street axes. The main gate on this side appears to have been flanked by two towers. Both ground-plans seem to indicate the presence of an earlier wall beneath. Bechert (1971, 258–9) has attributed an alteration to the gateway to the last quarter of the 2nd century. For the moment the question of later alterations to the gateways has to be left with the hope that the present investigation of the so-called 'Burginatium Gate' (the main gateway in the northern wall) will provide a more satisfactory answer.

The colonial wall: conclusions

Drawing the threads together, it can be stated that the erection of the town wall was begun on the river frontage at the north-east corner tower, in the spring of AD 106 shortly after the founding of the colony (see Fig 18). The founding of the colony, for which there has previously been only a *terminus post quem* of AD 98 and a *terminus ante quem* of AD 105, must now be coupled with the transfer of the Tenth Legion from Nijmegen to Pannonia in AD 104; the community which already existed at Xanten (see below) was raised to the status of a colony in order to settle the veterans and those soldiers shortly to retire, as well as to revitalize the legion in bringing it up to strength with recruits before sending it on its way to Pannonia.

The construction of the wall proceeded southwards and each section was presumably erected while work on the

Fig 32 The Mole Tower: remains of the pile grating beneath the foundation

towers was in progress; but the construction must soon have ground to a halt, possibly owing to lack of building material. At a time yet to be determined the work was restarted, but in a modified form, and the wall was completed in a more economic fashion. The way in which the gates and towers project only slightly from the wall suggests that the scheme was completed in the first half of the 2nd century. The design of the defensive circuit suggests it was planned by a military architect, who is assumed to have belonged to the ranks of the Tenth Legion or its veterans. The wall enclosed an area of some 73 ha, was 3400 m in length, and probably incorporated three twin-towered gateways, at least two single-towered gates, and a further nineteen towers (Fig 18). Approximately 52,000 cubic metres of stone and cement must have been employed in its construction: the stone had to be transported from the Eifel region 150–200 km distant by barge along the Moselle and Rhine.

The other defensive systems

Throughout the investigations on the northern and eastern sides it could be seen that the wall had cut through earlier features belonging to the 1st century settlement which was partially submerged by a tributary of the Rhine. To the west of the eastern wall were found further 1st century features which showed that the settlement was of the same extent as the later colony. The domestic architecture seems to show mediterranean affinities, ie wooden-framed houses with porticoes, and the alignment of the street grid was adopted by that of the colony.

The northern limit of this settlement has not yet been found but about 100 m west of the *cardo maximus* and not too distant from the Burginatium Gate Rüger located a rampart of earth and timber which possibly belonged to the settlement and marked its western boundary. North of the south-east corner tower a V-shaped ditch has recently been found, which is probably part of a double ditch system found during excavations in the amphitheatre in 1936 (Heidenreich 1940, Tafel 6). According to the latest finds the ditches had already been filled in by the end of the second half of the 1st century. To the south of these ditches no early settlement features have been seen. It is therefore possible that the settlement founded at the beginning of the 1st century was defended by twin ditches and a palisade bank, as found by Rüger in the north-western part of the town (Rüger 1980).

The character of this settlement, whether of a civil or military origin, must remain an open question until structures from this period have been examined. Bearing in mind that a not inconsiderable part of the settlement's eastern side was submerged by a tributary of the Rhine, it might well have been as large as the later colony itself. Several destruction levels can be assigned to the period preceding the beginning of the 2nd century when Trajan under the *ius Italicum* founded the colony (which was situated slightly to the south and west of the earlier settlement).

To conclude, mention must be made of a third defensive system which has come to light (the results of recent excavations have been brought together in Rüger 1979, 499–524). Two parallel ditches in the western part of the colony had been sectioned by Hinz and Binding and were clearly neither of medieval origin nor tank traps from the

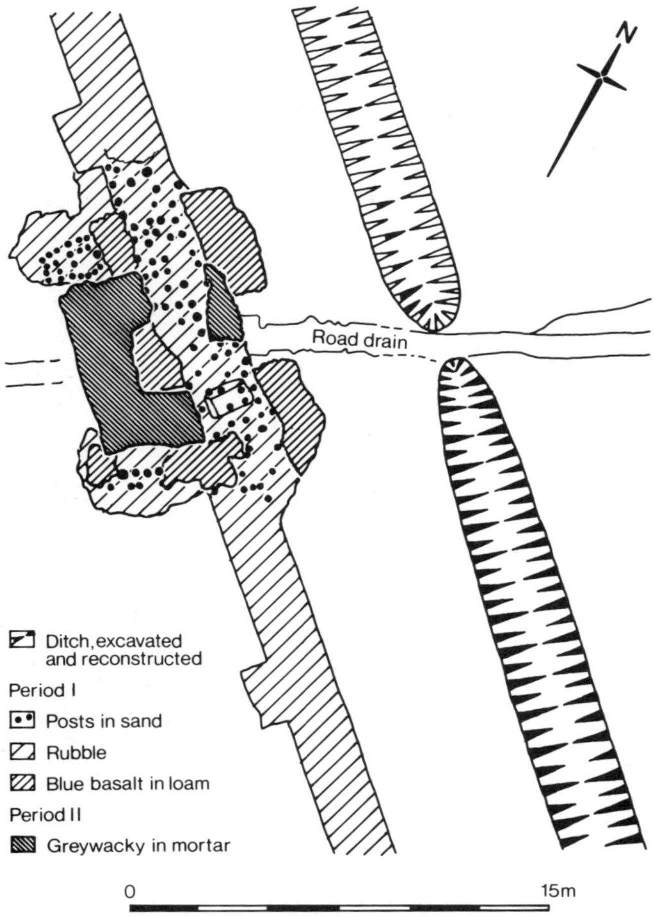

Fig 33 Small Harbour Gate, ground plan

last war. An excavation begun in 1974 with the aim of determining the southern boundary and date of this defensive system provided further details, and investigations in the last few years examined the northern and eastern sides, after Scollar had ascertained the course of the double ditch using the magnetic prospection method.

The following picture has emerged (Rüger 1979, 499–515): a fortified wall, with turrets at short intervals and fronted by a double ditch, was built around the inner nine insulae of the colony enclosing an area of sixteen ha. The wall itself had been totally destroyed by medieval robbing, leaving the foundation trench as the only evidence of its course. This was 4·50 m wide in the north-west but the wall is thought to have been only some 1·50 m thick at foundation level.

The curved foundation trenches of regularly spaced features represented a chain of semicircular towers between 7·50 m and 8 m in diameter which projected up to 5 m beyond the wall itself and were spaced at intervals of approximately 32 m (Fig 34). A circular bastion was found at the south-west corner and the north-east corner bastion was discovered in 1978 while the colony's street plan was being investigated. Rüger concludes (1979, 505) that a wooden bridge was laid out on the axis of the *decumanus maximus* over the double ditch, leading towards a gateway. On the eastern side there was a squarish robber-trench which Rüger interprets (1979, Abb 15) as a rectangular tower, part of a gateway flanked by two bastions. However, it is uncertain whether it actually was a squarish tower. The discovery amongst the destruction debris of a massive stone block on which scratch-marks were clearly visible suggests that the colony's street system continued to be used and that a gateway would have stood here, but the supposed tower's position directly on the axis of the *decumanus maximus* would seem to counter the argument that it belonged to a twin-towered gate.

The two parallel ditches were separated from the wall by a 6 m wide berm, and were approximately 12 m wide and 3 m deep. Many urban features of the colony were destroyed by the digging of the ditches, which seem to have lain open for a considerable time and were only filled in during the post-Roman period, before eventually disappearing under plough soil. No information as to the date or duration of this later Roman defensive system could be gained from the evidence of the finds. Occasional sherds of 4th century date were found but only in contaminated levels of the fill.

All that is certain is that the ditches, wall, and bastions must be later than the colonial structures of the 2nd and 3rd centuries. However, from an analysis of numismatic evidence based upon coins found (a) within the wall circuit of the colony but outside the later defence work and (b) within the defence work itself, Rüger concludes (1979, 515–24) that the later defensive circuit was laid out at the time of Constantine's building programme at the beginning of the 4th century. He also places its demise (in the light of the drastic reduction of the number of coins in circulation) in the period of the usurper Magnentius. As the late mintings of Magnentius are not found, the fall of the fortification at Xanten can be linked with the defeat of the Gallic usurper by Constantius II at Mursa in Illyricum (28 September 351) resulting in the reduction of the strength of the troops on the Rhine *limes*.

Postscript

This paper was written in March 1980. Since then further investigations have been carried out within CUT which have supplemented or corrected the facts presented above. For instance, it has been possible to examine the main

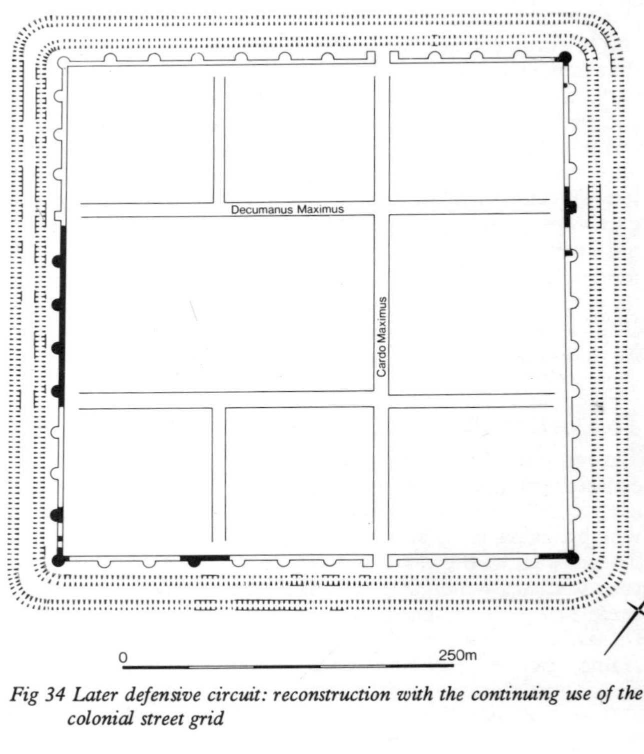

Fig 34 Later defensive circuit: reconstruction with the continuing use of the colonial street grid

drain on the northern side of the colony in front of the northern gateway (the so-called Burginatium Gate). Its sides consisted of greywacky masonry, and not of tufa blocks as had been thought previously, and must have been arched over, probably to the internal height of some 1·8 m with an internal width of 0·80 m. Because of later disturbance, which had reached the top of the drain, it could not be ascertained whether a ditch had been constructed above it or not. Another ditch does not seem to have been present on the northern side.

The foundation levels of the Burginatium Gate have also been investigated. Again they consist of a gateway flanked by two towers which project outward, one 2·3 m, the other 2·9 m. The gateway has the unusual feature of having been constructed over a cast-concrete platform, while the gate towers were erected on hollow foundations. The whole structure is some 10·70 × 27·80 m in plan and set upon a bed of piles.

Further evidence in connection with the system of twin ditches, which runs directly through the later amphitheatre, has apparently shown that it cannot belong to the earth and timber rampart discovered by Rüger south of the Burginatium Gate. In front of the eastern wall a corner of this system was found where it turned to the south. One may say, with some degree of certainty, that we have touched upon a previously unknown defence work lying south of the later colony (cf Gerlach 1982).

Acknowledgement

The text was translated from the German by C J Bridger, who is a member of the excavation team in the Archaeological Park.

References

Bechert, T, 1971 Römische Lagertore und ihre Bauinschriften, *Bonner Jahrb*, **171**, 201–87

Gerlach, G, 1982 Die Grabungen im Südosten der CUT, *ibid*, **182**, 325ff

Heidenreich, K, 1940 Das Amphitheater der Colonia Ulpia Traiana bei Xanten, ein Versuch seiner Wiederherstellung, *ibid*, **145**, 33–62

Precht, G, 1975 Der Archäologische Park in Xanten, Kreis Wesel, *Rheinische Ausgrabungen '75*, Sonderheft Januar 1976, 55–63, Das Rheinische Landesmuseum Bonn

——, 1976 Der Archäologische Park Xanten, Kreis Wesel, *Rheinische Ausgrabungen '76*, Sonderheft Januar 1977, 67–78, Das Rheinische Landesmuseum Bonn

——, 1977 Der Archäologische Park Xanten, Kreis Wesel, *Ausgrabungen im Rheinland '77*, Sonderheft August 1978, 156–81, Das Rheinische Landesmuseum Bonn

——, 1978 Der Archäologische Park Xanten, Kreis Wesel, *Ausgrabungen im Rheinland '78*, Sonderheft Januar 1979, 132–47, Das Rheinische Landesmuseum Bonn

Precht, G, & Rupprecht, G, 1976 Xanten, Colonia Ulpia Traiana, Bericht über die Forschungen vom 1.4.1974 bis 28.2.1975, *Bonner Jahrb*, **176**, 337–56

Rüger, C B, 1979 Die spätrömische Grossfestung in der Colonia Ulpia Traiana mit Beiträgen von G Binias, M Gechter und V Zedelius, *ibid*, **179**, 499–524

——, 1980 Research on the *limes* of Germania Inferior (German part) 1974–1979, in *Roman frontier studies 1979* (eds W S Hanson & L J F Keppie), Brit Archaeol Rep, **S71**, 495–514

Scollar, I, 1965 *Archäologie aus der Luft*, Düsseldorf

Steiner, P, 1903 in *Bonner Jahrb*, **110**, 182–7

——, 1911 *Xanten-Kataloge, West-und Süddeutsche Altertumssammlungen*, **1**

Stoll, H, 1936 in *Germania*, **20**, 184 (see also Oelmann, F, 1936, *Bonner Jahrb*, **140/41**, 431)

von Petrikovits, H, 1937 Xanten, *Bonner Jahrb*, **142**, 248–9

——, 1952 Die Ausgrabungen in der Colonia Ulpia Traiana bei Xanten—Die Ausgrabung der Kernsiedlung und der Uferanlagen (1934–1936), *ibid*, **152**, 43–161

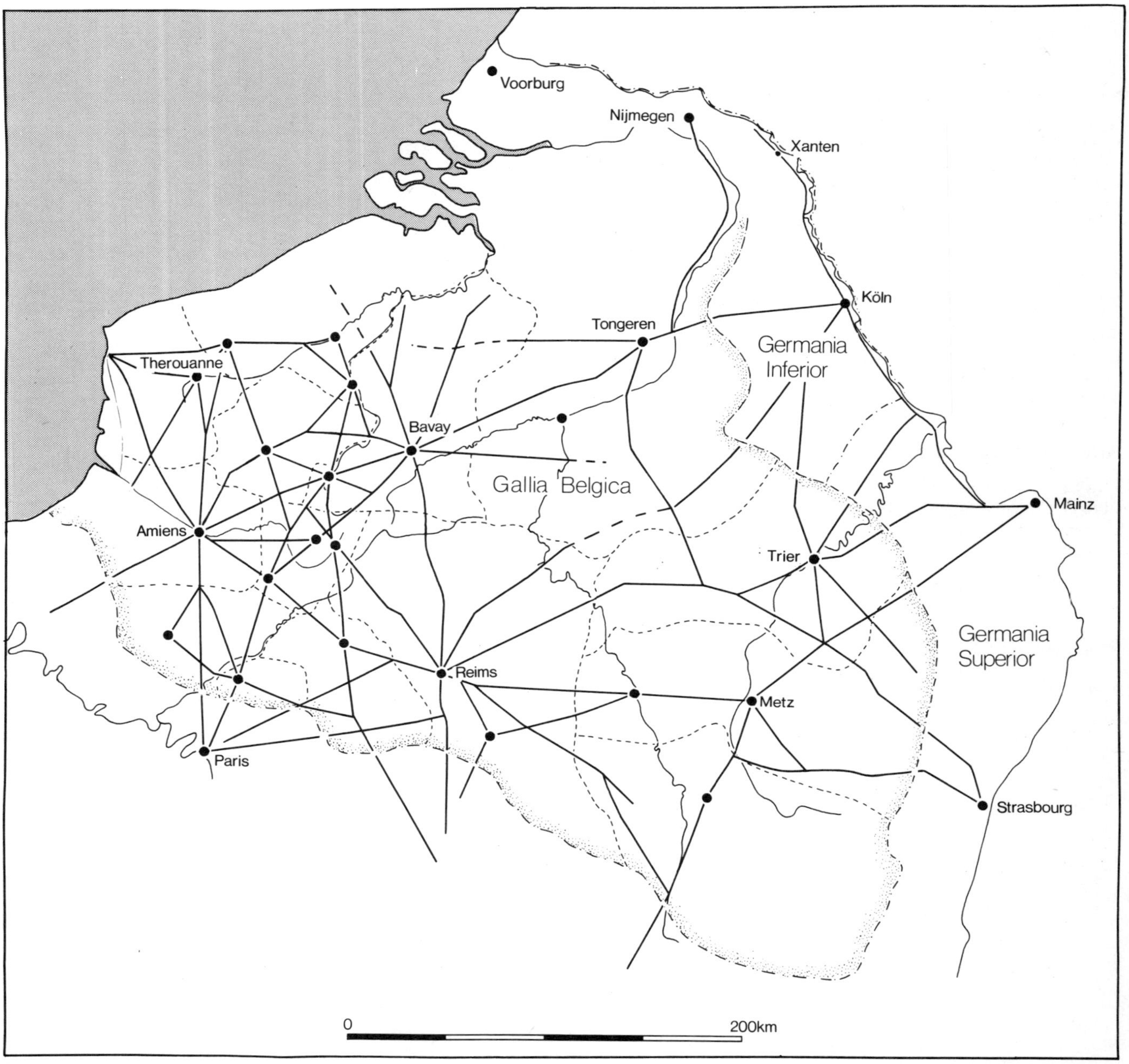

Fig IIIa Gallia Belgica in the early Roman period

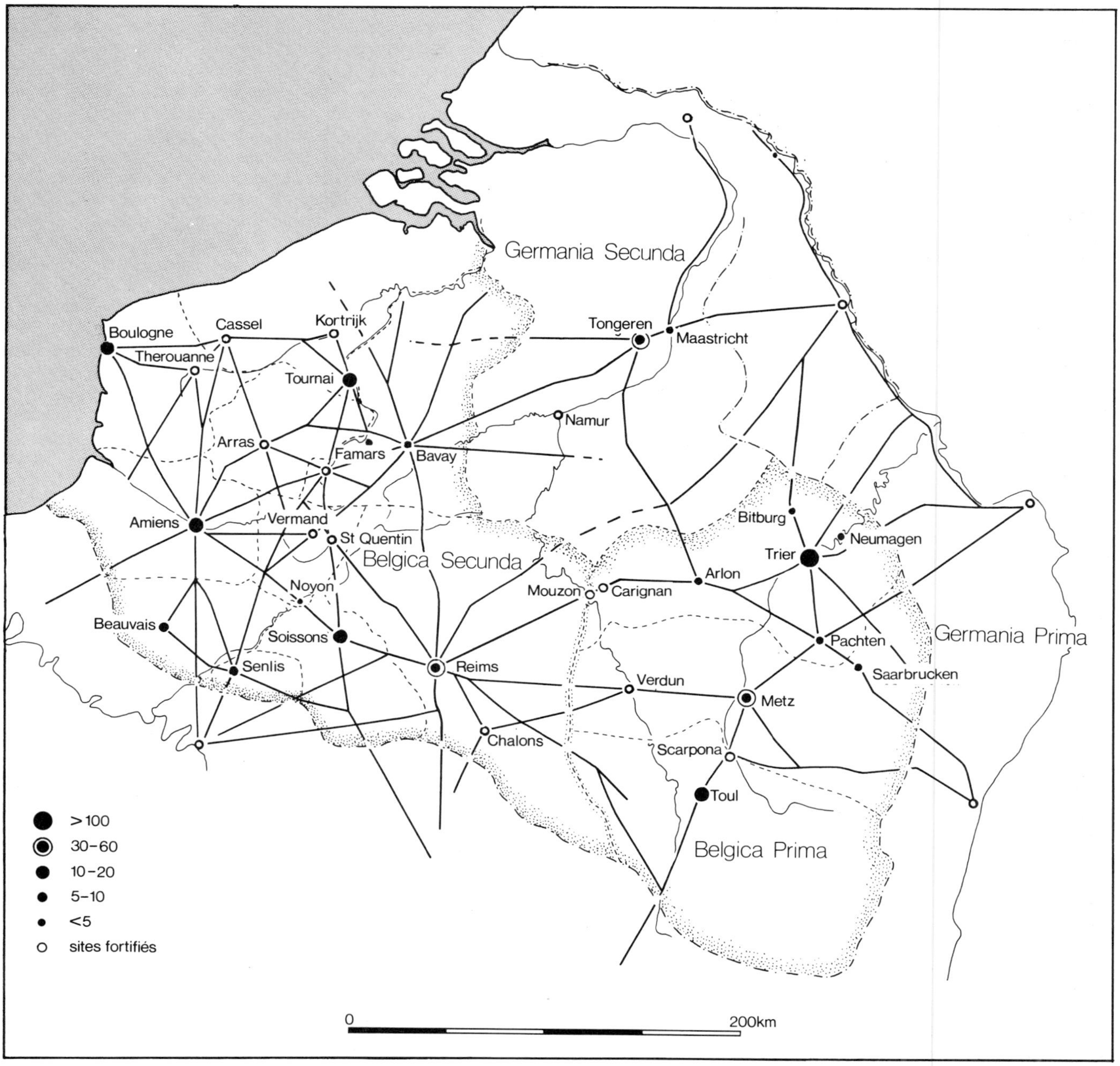

Fig IIIb Gallia Belgica in the late Roman period

In an article published early this century, but which still retains its value, Cumont (1914–19, 81) stated that Belgium in the Roman period 'where such important communities were later to develop, was a country without towns'. This statement may appear exaggerated, but from the Roman point of view it was virtually true. As far as Rome was concerned, any settlement which had not achieved the status of *colonia* or *municipium* was not a town. Northern Gaul was particularly devoid of communities of such status. To find them, one has to go to Germany or Holland, where Colonia Claudia Agrippinensium (Köln), Colonia Ulpia Traiana (Xanten), Municipium Batavorum (Nijmegen), and Forum Hadriani–Municipium Cananefatium (near Voorburg) all lay in the militarized zone next to the *limes*. Only two towns in Gallia Belgica received the title of colony: they were Colonia Treverorum (Trier) and Colonia Morinorum (Thérouanne). This is a surprisingly small total for a whole province. The absence of towns, in the legal sense of the term, was however a characteristic of northern Gaul, and creates a clear distinction between the situation there and in Germany or Britain; a map recently published by Hassel and Künzl (1975) shows this most clearly. If towns are seen in the forefront of the process of romanization, one must admit that the military zones were especially favoured, and that the army played a role in romanizing these regions. Archaeology confirms, however, that as well as these chartered towns, Rome created other urban centres in the north. Several of them were 'proper' towns, with a rectilinear plan, a forum, temples, an amphitheatre, a basilica, or baths. One finds all the elements which signify the normal development of town life. With few exceptions—possibly Rheims, for example—they were new towns, planted not on pre-existing native settlement sites, but at a distance from them, normally at the crossing-points of the newly established road system. None of them succeeded a fully-fledged legionary fortress: at most, some of them grew from a partly military and partly civilian establishment of Augustan date. One is left with the impression that Rome wished to leave the Gallic system of tribal areas largely undisturbed. It was the *caput civitatis* which formed the political, administrative, and religious centre of a region. The subject of the origin of towns in Gallia Belgica is a problem I will leave here (see in general, Clavel & Lévèque 1971, 43–5; for a bibliography, Raepsaet-Charlier 1975; Stillwell *et al* 1976).

Rome did not treat all these urban centres on an equal footing. Some received the title of *colonia*, some *municipium*, others remained simply *vici*. The colonies alone merited a wall-circuit which was more often symbolic rather than a strategic necessity (Fig IIIa). This development is in any case a very late one in Gallia Belgica, where the earliest circuits date only from the 2nd century. These go in tandem with grants of colonial or municipal status to towns in the northern part of Germany.

Following the fundamental work of Blanchet (1907), a study was made of these wall-circuits twenty years ago (Butler 1959, 26–7) which distinguished the hallmarks of earlier and later Roman ramparts. Recent excavations, and the interpretation of some historical source material, have permitted some further refinements of the accepted picture. For the earlier Roman period, the cases of Trier, Tongres, Tournai, Thérouanne, and Rheims will be briefly dealt with here.

Trier

The origin and status of this town have been recently discussed by Ternes (1978) and Wightman (1970, 39–43; 1976). Founded on a terrace dominating the Moselle, the settlement reached a high degree of urbanization in the Augustan period. Probably under Claudius it obtained the mainly honorific title of colony. The town was the seat of the imperial fiscus, and an operational base for the armies of the Rhine. It was only later that the settlement was enclosed by walls. Discussion has centred for a long time on the construction date of the defences, and dates from the 1st to the 4th century have been proposed for the best-known element within the fortifications, the Porta Nigra (Lugli 1960, 97–101; Crema 1959, 558; Wightman 1970, 96). Detailed study of its architectural characteristics,[1] and of its archaeological context, has now given precision to this time-scale; the gate, as well as the ramparts with which it was associated, must date from the last quarter of the 2nd century (Gose *et al*, 1969, 86; Zahn 1977; Cüppers 1969, 170–2; 1973; 1977) (Fig 35). The total length of the circuit was 6418 m, enclosing 285 ha. Surviving walls are 3 m thick, and their foundations, nearly 4 m wide, are in places as deep as 3·50 m. The wall facing, which encases a core of stones and mortar, is of small limestone blocks, regularly squared and coursed, and set in a light mortar. The joints are finished in trowel-struck pointing of a yellowish colour (Fig 36). At about 0·80 m above ground level, four of the facing courses of the wall

Fig 35 Relief representing a Roman town gate (photo: German Archaeological Institute, Rome)

Fig 36 Trier: detail of the town wall-facing (after Trierer Zeitschr, **36** *(1973), 7)*

form a slight batter. The relatively infrequent towers, of which thirteen are known, are spaced irregularly, but normally on the axis of the street grid. They are massive, hollow, and circular, and their external diameter varies from 8·50 to 10·45 m, while internally their diameter is from 4·5 to 5·24 m. To the north and east, it has been possible to section the ditches protecting the wall: the inner ditch was 9 m broad, and 2·5 m deep; the outer one was 12/14 m broad, and 3 m deep (Fig 37). On the southern side, a third ditch, 14 m broad and 4 to 5 m deep, completed the defences. The gateways incorporated within the walls were contemporaneous; they were constructed of large stone blocks, and had a double carriageway, flanked by semicircular towers. The construction of this circuit must have represented a considerable expenditure of effort. One should perhaps attribute it, not to the threat of danger, but to the desire of the citizens to embellish their town with ramparts and monumental gateways.[2]

Tongres

Although it was the *caput* of the *civitas Tungrorum*, the town never achieved urban status, in the Roman legal sense (Fig 38). The settlement was not the successor of a native site, nor of the famous Caesarian camp of Atuatuca, despite the survival of the name. To the west

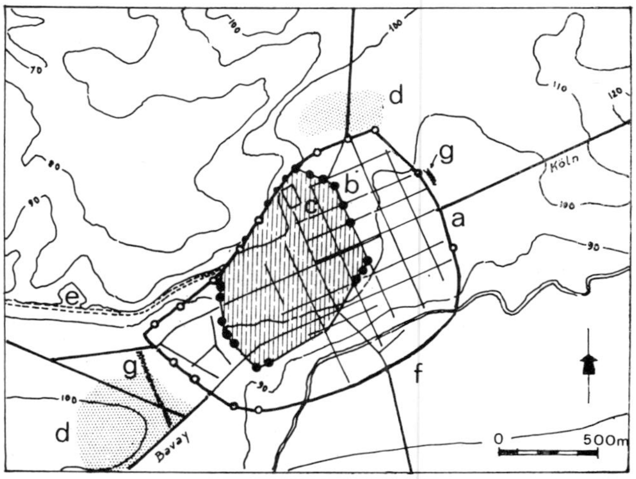

Fig 38 Plan of the town of Tongres: a first wall-circuit; b second wall-circuit; c temple; d cemeteries; e aqueduct; f wall on piled foundation; g Augustan ditches

of the town, excavation has revealed traces of V-shaped ditches and a timber rampart, dating from the Augustan period (Fig 39). A similar ditch has been sectioned on the eastern side of the town, and it is possible that an area of 140 ha was enclosed within the Augustan defences. These do not seem to have lasted long: the ditches were filled with clean clay without traces of silting. In the centre of the settlement, but not related to the pattern of the street grid, a cremation tomb of Tiberian date has been located (Vanvinckenroye 1964, 19–20; 1975, 17–22). This arrangement is reminiscent of discoveries in certain British towns, for example Silchester or Verulamium. The regular street grid, which covered the eastern portion of the town, did not succeed the regular layout of a pre-existing legionary fortress (Ulrix 1962/63), but dates from the Claudian period. Excavation has located few public buildings; one, in the northern sector, was a temple of native type (Mertens 1967).

The whole built-up area was surrounded by a vast circuit of walls, 4544 m long, enclosing an area of about 136 ha. These defences, traditionally dated to the Claudian period (de Schaetzen 1955), have been established, principally by work on the northern and western sides, to belong to the second quarter of the 2nd century (Vanvinckenroye 1975, 37–8) (Fig 40). The walls, 2·15 m broad, were very carefully built. The foundation was usually slightly broader than the superstructure of the wall, and was composed of stones laid slantwise in clay. It was of variable depth, between 0·60 and 1·20 m, and at the top was finished off by several courses of rather larger coarse stones bonded together with mortar (Fig 41). In

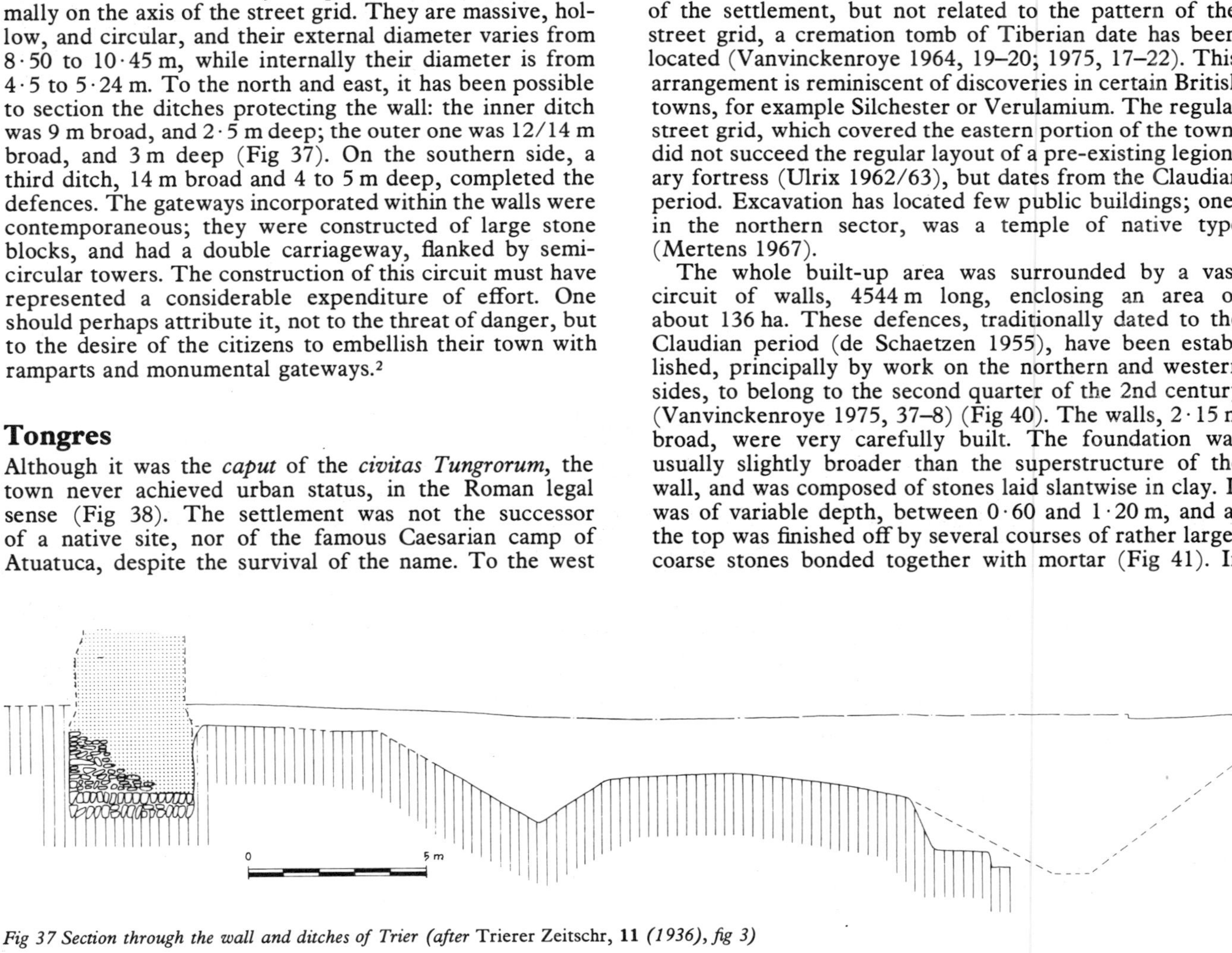

Fig 37 Section through the wall and ditches of Trier (after Trierer Zeitschr, **11** *(1936), fig 3)*

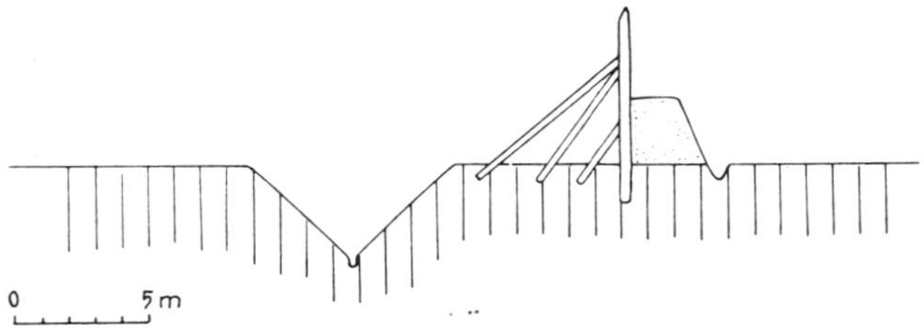

Fig 39 Tongres: section of Augustan ditch and palisade (after Vanvinckenroye 1975, fig 4)

elevation, the core consisted of a series of horizontal courses made of a mixture of flint and solid yellowish mortar. The inner and outer facings were of small blocks of sandstone or flint, regularly squared, and measuring 90–110 mm by 150–180mm (Fig 42). On both sides, the wall below ground level was stepped out in several offset courses of 0·80 m breadth. At 0·34 m above the last offset there was a batter 0·41 m high formed from three of the normal courses (Fig 43).

Fifteen circular towers were set at irregular intervals along the circuit, often on the axial lines of the streets. One of them, probably representative of all, had an internal diameter of 5·36 m, which externally was 9·04 m. On the southern side of the town, where the subsoil is particularly marshy, the wall was only 1·40 m thick. It was built on a foundation of large regular limestone blocks, 1·4 m long, 0·70 m wide, and 0·60 m thick, alternating as headers and stretchers. When used as stretchers, the blocks were placed together dry-jointed (Fig 44).

This foundation rested on a network of oak piles driven vertically into the ground (Fig 45). They were between 0·25 and 0·30 m square, and were nearly 5 m long. They were placed in rows of four abreast, and three such rows (twelve piles) were used for each linear metre of wall (Fig 46). The construction of at least 700 m of the wall in this

Fig 41 Tongres: detail of the first wall-circuit (copyright ACL Brussels)

Fig 40 Tongres: the first wall-circuit

Fig 42 Tongres: detail of the wall-facing of the first wall-circuit

Fig 43 Tongres: a tower of the first wall-circuit (1) with the later Roman wall (2) leaning against it

Fig 44 Tongres: detail of the wall in the marshy area (copyright ACL Brussels)

Fig 45 Tongres: the wooden piles under the early rampart (copyright ACL Brussels)

style would have demanded the felling of nearly 8500 trees. Over the top of these piles, there was a thick mortar layer, composed of gravel, chalk, and flint, and intended to take the large foundation blocks. On this base, the wall was constructed. The few repairs and modifications noted do not seem to have been carried out substantially later than the construction date (Fig 47).

All the gates, situated at road junctions, had a monumental aspect. One, the West Gate, could be completely excavated (Mertens 1964, 22) (Fig 48). It was 20 m wide, and had two carriageways each 3 m wide flanked by rectangular towers measuring 6 by 10 m. In the mortar on top of the foundations the marks of large stone blocks were still visible (Fig 49). This leads one to suppose that, like the Porta Nigra at Trier, this gate too had a monumental aspect.

The circuit was defended throughout its length, except in the marshy area, by two or three ditches. A section to the west of the town revealed their profiles. They were V-shaped, and regularly cut into the clay subsoil (Fig 50). The point of the V was finished off in all cases with a small rectangular slot, 0·27–0·29 m in width. The innermost ditch followed the course of the wall exactly, even bending round the circular towers (Fig 51); it was 6 m broad, 3 m deep, and its centre lay at 5·85 m from the wall. The middle ditch was 17·80 m from the wall and was 14·5 m broad, and a third, outer ditch parallel to the others lay at 33·25 m from the wall, and was 16 m broad. It is possible that the outer ditch was not contemporaneous with the two others. The slope of the ditch sides may indicate the

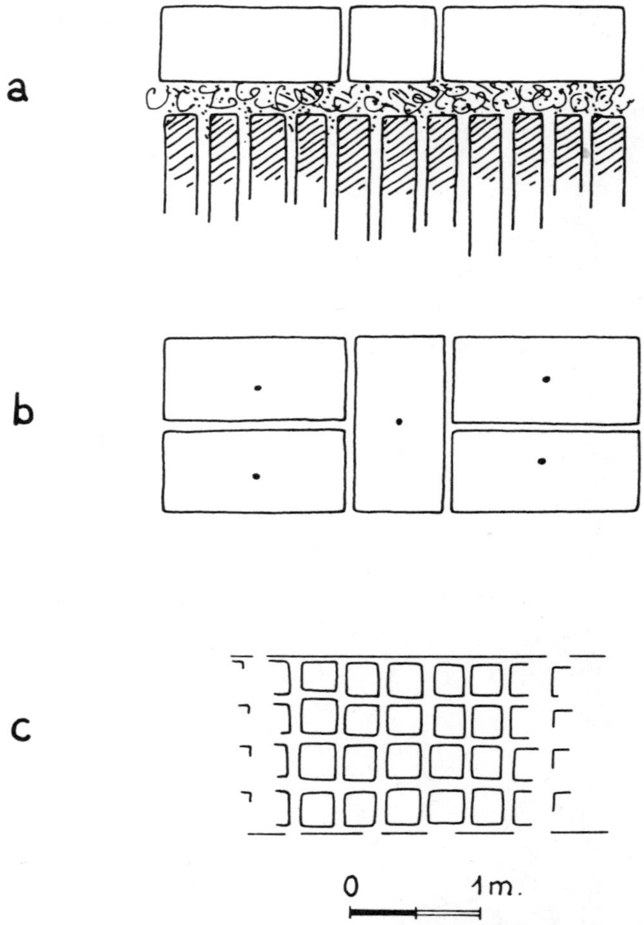

Fig 46 Tongres: sketch to show the layout of stone and wooden piles under the wall-circuit

Fig 47 Tongres: one of the towers of the early wall-circuit

height of the parapet, which therefore could have reached 6 or 7 m (Mertens 1964, 20–3) (Fig 52).

So far, research has indicated that Trier and Tongres were the only towns in Gallia Belgica to have had defensive circuits in the early imperial period. Other towns are sometimes claimed to have been defended at the same time and we shall examine briefly the cases of three commonly cited cases, Thérouanne, Rheims, and Tournai.

Thérouanne

Thérouanne, whose Roman name was Tervanna, is recorded as a colony in an inscription discovered in the 15th century near Nijmegen (*CIL* XIII, 8727), but the title appears to have been purely honorific. Despite the high degree of urbanization and the town's importance, it appears never to have been enclosed with walls (Delmaire 1976, 133–9; Bernard 1975).

Rheims

At Rheims (Durocortorum), the seat of the governor of Belgica, the town developed during and subsequent to the

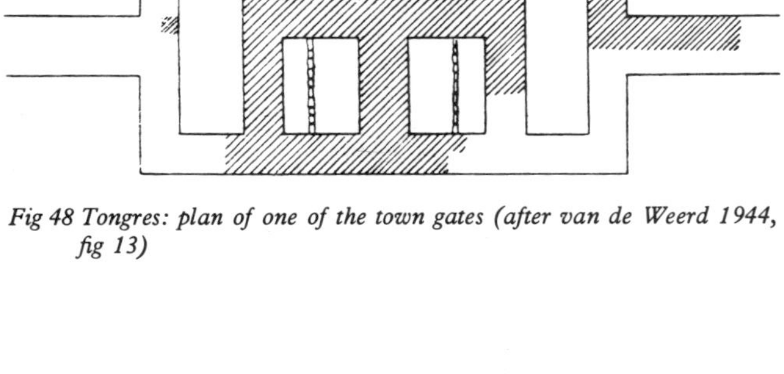

Fig 48 Tongres: plan of one of the town gates (after van de Weerd 1944, fig 13)

Fig 49 Tongres: detailed view of the foundation of the western gate (copyright ACL Brussels)

Fig 50 Tongres: view of the defensive ditches west of the town

Flavian period. It had a rectilinear plan and several monumental buildings, among them a forum with a *cryptoporticus*. In several reports made at the time of chance discoveries or hastily mounted excavations, reference is made to one or two ramparts dating to its pre-Roman Gallic period (Fig 53). The outer rampart would have enclosed the whole of the Roman town, while the inner rampart coincides in a surprising fashion with the circuit of late Roman defences. Several vague traces of the outer rampart have been located and sectioned, but their date has not been established (Frézouls in Stillwell *et al* 1976, 288–9; Hollande 1976). The four 'town gates' did not lie on this circuit: they were incorporated within the later one.

Tournai

Relying on an old interpretation of the evidence (Amand 1951, 384–9), it has been supposed that there was a walled circuit at Tournai (Butler 1959, 26). Up until now, excavation has not confirmed this hypothesis.

In view of all this, the results of our search for town defences of the earlier Roman period in Gallia Belgica might appear disappointingly small. Only two towns had defences; the difference between this and the provinces of Germany or Britain is remarkable, and calls for an explanation. Why were town defences built? The clearest reason

Fig 51 Tongres: the ditch and the early wall-circuit

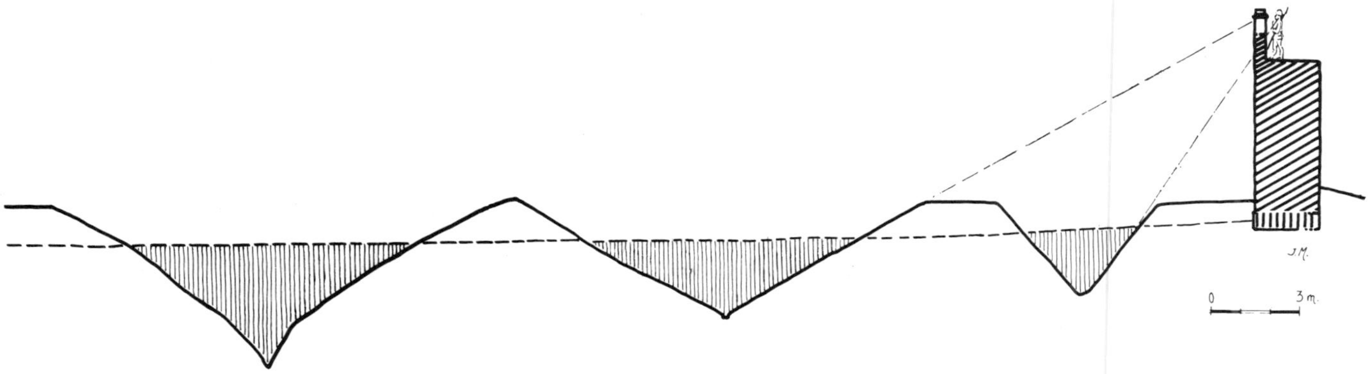

Fig 52 Tongres: profiles of the ditches (after Mertens 1964, fig 10)

would seem to be the presence of some danger, and in the northern part of Gaul almost all the fortified towns lay not far from the frontier. However, all these defences, except that of Cologne, were built under the prevailing *pax Romana.* Should one interpret them, therefore, as the desire of the townspeople to show off their wealth? Other towns, lying deeper within the province, had the same economic development without feeling the need to build walls. Should one see the walls instead as having value as propaganda? Did the Romans wish to lay out before the barbarian world the cities in all their splendour and technical perfection, the decorative monuments, the imposing defences, and, in a word, dazzle them with the beauty and the attractions of their civilization? There may have been other factors—for example, the status of these towns. For almost all the colonies, a wall-circuit was a symbol of an

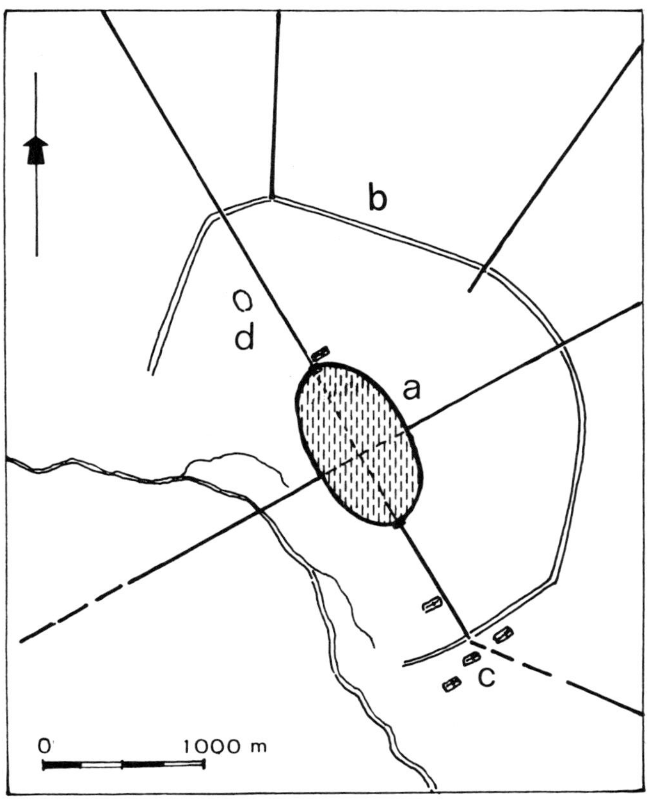

Fig 53 Plan of the wall-circuits of Rheims (after Neiss 1976): a the late Roman town; b pre-Roman ditch(?); c the St Rémy quarter; d amphitheatre

entitlement to their noble status (Harmand 1961). Or is there a chronological element, with colonies and fortified towns becoming less frequent foundations in the 2nd century than around the beginning of the era?

Whatever the answers, these early Roman wall-circuits have several characteristics in common, as Butler pointed out. They enclosed almost the whole of extensive areas of settlement, they were protected by multiple ditches, the towers were few in number, and the gates had a monumental and decorative character. These walls were the privilege accorded to only a few towns. Other settlements, like Amiens, Bavai, Metz, Senlis, Boulogne, or Rheims, had a highly developed urban layout, but had to wait until the end of the 3rd century before receiving the protection of defensive walls. And in the late Roman period, the picture changes completely.

The insecurity of the period provoked a rash of fortifications constructed round urban centres, or intended to protect land and sea communications by guarding the main routes. Even rural communities sheltered under the protection of walls, while the countryside itself was dotted with refuges and other fortifications (Mertens 1980; Clavel & Lévèque 1971, 69–71). These fortifications have some characteristics in common, but their siting, form, and function are extremely diverse. Several studies have been devoted to these in recent years, and the paper by Johnson in this volume deals more widely with them. What I have to say here is limited to several general points, and to some new discoveries revealed by recent excavations, about some of the walls in Gallia Belgica.

By this time defences were no longer merely symbolic, but a real necessity, taking more account of the actual importance of a site than of its official status. The defensive walls now had a strategic importance; they enclosed only the vital hearts of the towns. Ringed with many towers, as often as not they incorporated one or other important city monument. The area they enclosed was often so contracted, and their circumference so tightly drawn, that they appeared more like citadels than fortified towns.

The present state of research on towns in the area is summarized in Table 1. This gives the length of a town's wall-circuit, and the area enclosed, as far as can be determined.

The difference between these figures and those of the earlier period (see Figs IIIa & b) is remarkable. Disregarding exceptional cases like Trier (an imperial capital) or Rheims, Metz, and Tongres (*civitas* capitals), the majority of the fortified sites enclosed between 5 and 15 ha. The smallest, from 1 to 4 ha, resemble military installations or *annona* stores, and this is the interpretation that one can put on a passage from Ausonius, describing the countryside of the River Moselle in the Trier region, when he

Table 1

Name	*Length of wall-circuit in metres (approx)*	*Area in hectares (approx)*
Trier	6418	285
Rheims	3200	60
Metz	3100	55
Tongres	2604	43
Tournai	1520	13
Boulogne	1460	12
Soissons	1400	12
Toul	1300	10
Amiens	1200	10
Beauvais	1170	9
Senlis	850	6·5
Arlon	780	4·9
Bavai	660	4 (western *castellum*, 2·5)
Maastricht	650	2
Pachten	570	1·9
Famars	547	1·8
Bitberg	570	1·8
Neumagen	395	1·5

called Neumagen '*non castra, sed horrea Belgis*' (not a fort, but a granary for the Belgae) (*Mosella*, line 457).

The plans of town walls have been mentioned above. The circular or oval shape was frequently used: this was because the highest areas of the town were enclosed by walls (as at Tongres, Senlis, or Bitburg) or because of the terrain itself (as at Arlon). A square or rectangular plan took account of the rectilinear grid of earlier days (Soissons or Amiens) or was the result of a purely military occupation (Boulogne?) (Gosselin *et al* 1976; Will 1960, 363–5).

In several places, these fortifications incorporated existing monuments within their circuit: at Amiens the amphitheatre served as a bastion (Vasselle & Will 1960; Vasselle 1962; Will 1962a, 79–84; Raepsaet-Charlier 1975, 170–2). The old forum at Bavai and its adjacent buildings (Will 1962b), the baths at Famars (Bersu & Unverzagt 1961; Brulet 1975, annexe 2, 64–85), the praetorium (?) at Tournai (Mertens & Remy 1974, 29–34), the northern temple at Tongres (Vanvinckenroye 1975, 44–7), and the monumental arches of the earlier town at Rheims, all formed part of the circuits.

From the technical point of view, there are certain similarities: the thickness of the walls, the use of large worked blocks from earlier monuments, and the frequent use of brick courses. But there are also differences: the tower shape (circular, U-shaped, square) or the wall-facing (number of brick-courses and the shape of the blockwork facing). Johnson (1973) had made a detailed study in a recent paper, where he concludes that building gangs and construction programmes may have existed. I think it is in this direction that research must progress; this approach would allow us to work out a chronological sequence for the construction work.

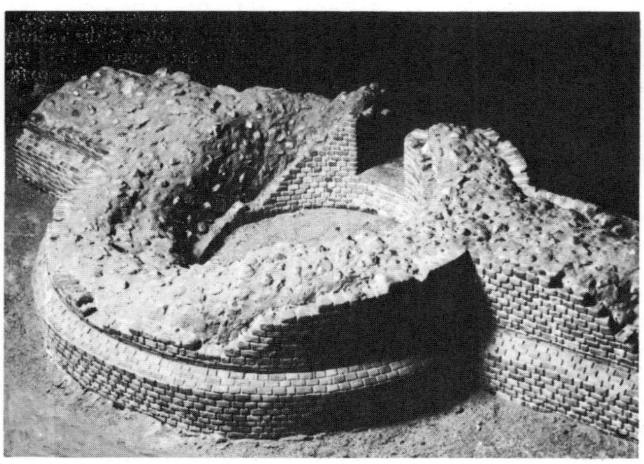

Fig 54 Tongres: one of the towers of the late Roman circuit, situated on the Vrijthof—model (photo: Museum of Tongres)

Fig 55 Tongres: detail of the masonry of the tower in Fig 54 (copyright ACL Brussels)

Many details are still missing. In order partially to fill these gaps, I will now record several new discoveries made about the walls of Tongres, Bavai, Famars, Tournai, Rheims, and Arlon.

Tongres

The late Roman wall at Tongres has been known for a long time (Fig 38). The width of the foundation varied from 3·60 to 3·80 m, and was composed of horizontal courses of blocks laid in sand, and covered with four courses of stone blocks (0·30–0·35 m thick), set in a bed of pinkish mortar (Fig 43). On the exterior face, an oblique batter about 0·25 m high was formed by three courses of blocks of tufa or limestone (Figs 54, 55). The superstructure of the wall had a width of 3·11 m, and was very carefully finished. The towers were circular (Figs 56, 57) with an external diameter of 8·06–9 m, and an internal one of 4·20 m. On the outside of the city wall a tower has survived to a height of 2·33 m, and on the inside up to 1·53 m. In the northern sector, where the subsoil was relatively unstable, one of the towers rested on a network of small wooden piles, 90×120 mm in section, and almost 1 m long (Fig 58). The towers were spaced at intervals of almost 25 m. An associated ditch, 9·50 m broad and 2·15 m deep, may have been recut in the northern sector (Mertens

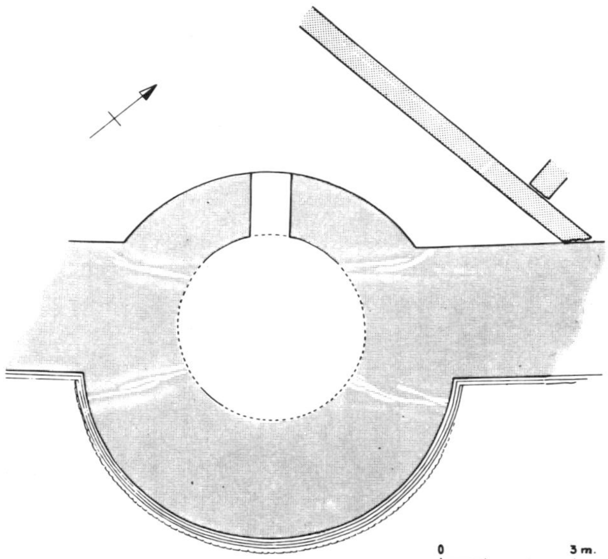

Fig 56 Tongres: plan of the tower in Fig 54

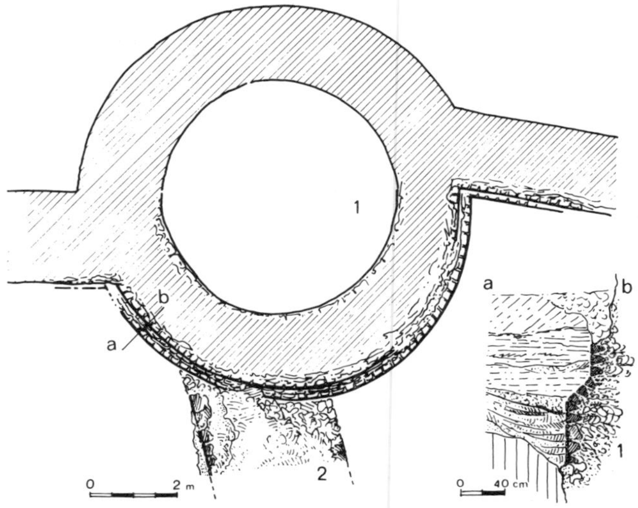

Fig 57 Tongres: a tower of the first wall-circuit (1) with the start of the late Roman circuit (2)

1977). In this same area, the late Roman walls incorporated and reused the early imperial circuit. Towers added to the earlier wall at this point were apparently semicircular.

Bavai

The walls of Bavai have been studied in detail (Will 1961; 1962b; 1963; Biévelet 1964; C Pietri in Stillwell *et al* 1976, 135–6). They enclosed a twinned pair of forts, and measure in total 400 × 100 m (Fig 59). Three phases of construction can be distinguished. In the first phase, the shops flanking the north and south of the monumental building (the *cryptoporticus* of the forum) were blocked by a wall 1 m thick, faced on both sides. The wall which marks the eastern side of the western *castellum* (Fig 60) may also belong to this original phase, although the method of construction was quite different. It was 3 m thick, rested on a foundation of large reused stone blocks, and was protected by semicircular projecting towers nearly 6 m in diameter. It is worth remarking that the mortar used does not correspond to that in the wall blocking off the shops. If this wall does indeed belong to the first phase, the fortress of Bavai would then have had only two towers, a small number for a late Roman wall-circuit.

The second phase is represented by a masonry wall with yellowish-white mortar which incorporated the whole of the two forts. Where it rested against the structures of the first phase, the wall only had an external facing. Elsewhere it was faced on both sides. It was about 3 m thick: the core was composed of broken stone and brick fragments and there were double tile courses 43 mm wide (with joints 35 mm thick) running through its complete width, at vertical intervals of between 0·82 and 1·50 m (Fig 61). On the northern and southern flanks, a single solid semicircular tower, with only a slight external projection, protected the wall. On the west, two towers, one solid, the other hollow—the only one in the whole circuit at Bavai—stood at the angles of the large monumental building, while a third tower stood on the axis. Their diameter varied from 7 to 11 m.

In a third phase of building, a new wall was butted up against the one just described. It was only faced externally and was compounded with pink mortar made from lime and crushed tile. Its thickness varied from 1·20 to 2·10 m (on the north, it measured 1·66 m). Many solid semicircular towers were built at intervals varying from 24 to 30 m. They encircled existing towers, were placed between them, or flanked the gates. In the two latter situations, they

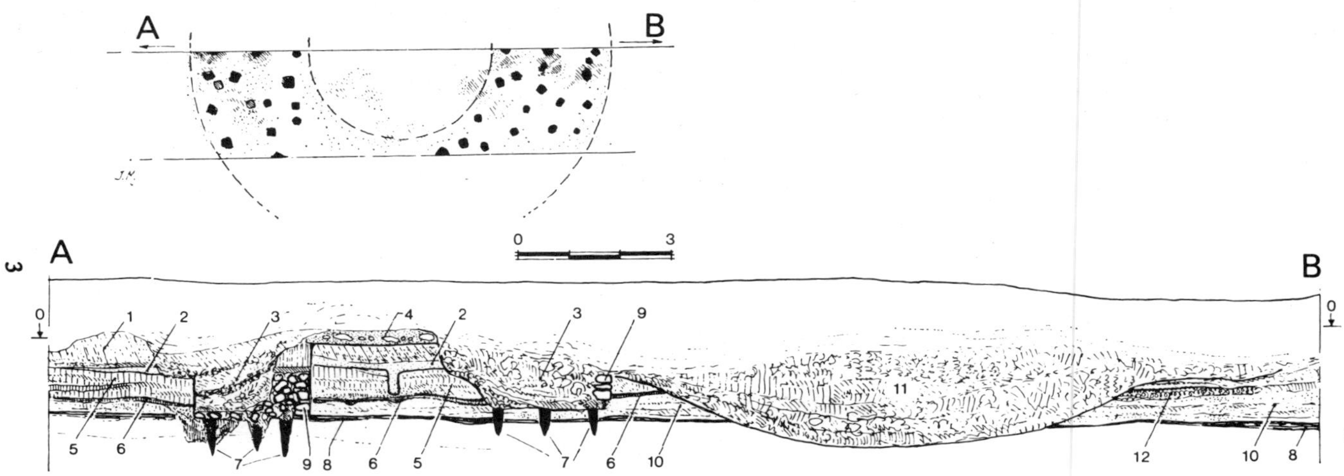

Fig 58 Tongres: plan and section of a tower of the late Roman circuit: 7 wooden piles under the tower; 11 ditch

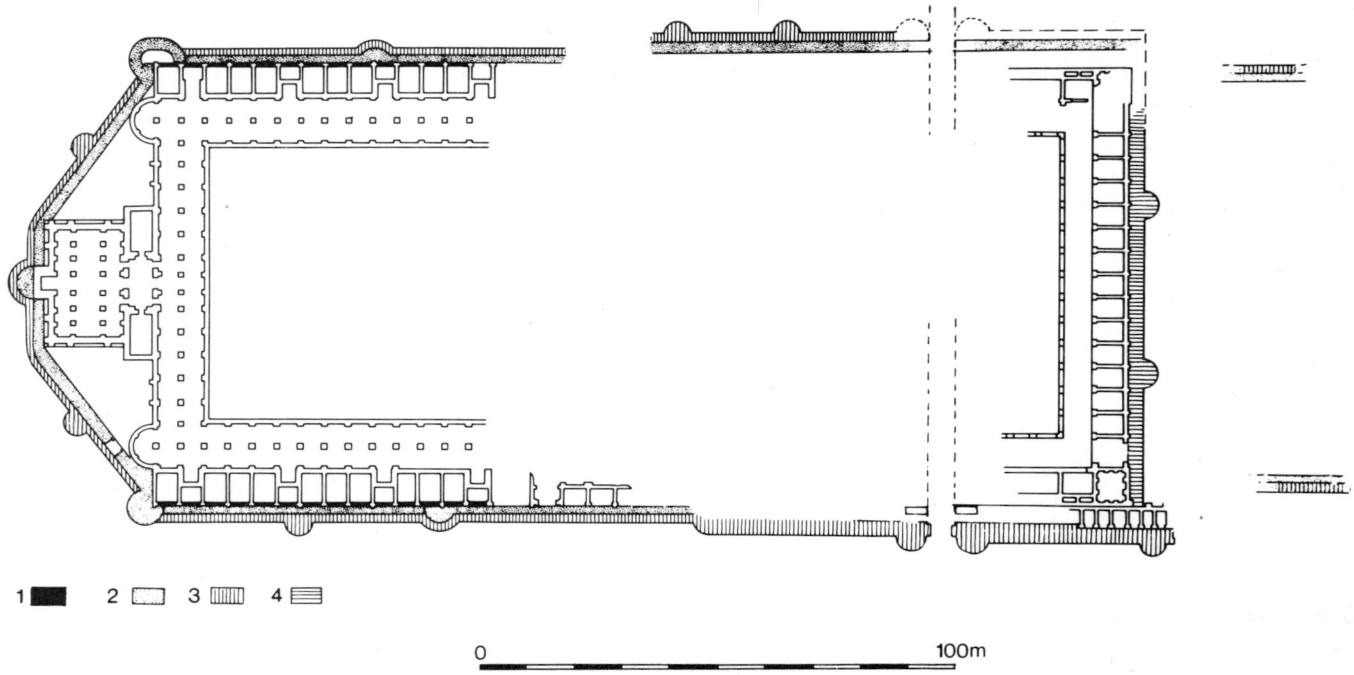

Fig 59 Bavai: overall plan of western castellum *(after Brulet 1975 and Boucly 1979, fig 10): 1 phase 1; 2 phase 2; 3 Valentinian period; 4 phase 2(?)*

Fig 60 Bavai: the east wall of the western castellum

Fig 61 Bavai: detail of the facing of the wall-circuit, phase 2, in the western castellum

generally had a diameter of a little more than 5 m. This wall, laid on a foundation of large blocks about 0·70–0·80 m high, had a facing of regular blocks of stone interspersed with a double course of tiles (0·43–0·57 m long and 0·42–0·67 m wide). The spacing of these tile-courses varied from 1·32 to 1·60 m. The rubble courses within the core were in many places set herringbone fashion. A ditch 8 m broad and 3 m deep has been located along the southern flank: it is difficult to determine to what phase of the defences it belonged. Further reference will be made to the chronology of these walls.

Famars

The *castellum* at Famars had an irregular quadrilateral plan, 105–110 m by 140–165 m. The western side was bowed outwards because it included the baths of the earlier Roman *vicus* within the circuit. On the eastern side, two phases of construction have been established: the first is represented by a wall 1·20 m thick, whose facings were made of double tile-courses alternated with three (sometimes only two) courses of regularly-shaped stone blocks (Bersu & Unverzagt 1961, figs 22, 23). Semicircular towers, 5·20 m in diameter, abutted the wall but were not bonded into it, despite having a similar facing. At the south-western angle, a semicircular tower of one build with the wall only projected internally: it actually formed a passage in this corner of the fort. In the second phase, the curtain wall was strengthened by the addition of another wall, 2·30 m thick, which also enclosed the tower. This second wall was bonded with pinkish mortar: no tile-courses have been found in it, and the ashlar facing is less carefully finished than in the earlier wall (Bersu & Unverzagt 1961, 173–4).

On the western side, the original wall was 1·80–2 m thick. The wall proper was built on a foundation of large reused blocks, and consisted of a core well-faced internally and externally. Tile-courses were constructed at various levels in the wall. Solid semicircular towers, 7 m in diameter, were built shortly afterwards against the wall. On its internal face, a series of postholes in the masonry provides an indication of the height of the wall-walk. The phases of construction can also be traced in the ditches dug round the fort: recent excavations to the west of the fort (Brulet 1975, annexe 2, 65–85) have identified three defensive ditches. The first, at about 10 m from the wall, was 3·25 m deep. The second was about 20 m away, and the third, 12 m broad and more than 4 m deep, has been found 32·50 m from the defences. There will be further discussion below of the chronological problems presented by these walls.

Tournai

The late Roman circuit of walls at Tournai was located during the course of excavations in 1955 (Fig 62), when a length of more than 85 m was discovered (Mertens & Remy 1974). The width of this particular length is still unknown, since the exposed portion abutted the wall of an earlier Roman building. The total thickness of both walls varied between 2·50 and 3·25 m. At the western and southern ends of the building, the wall was protected by two circular towers, 8–9 m in diameter (Fig 63). Elsewhere, the wall was 1·25 m wide, and the external facing was of regular

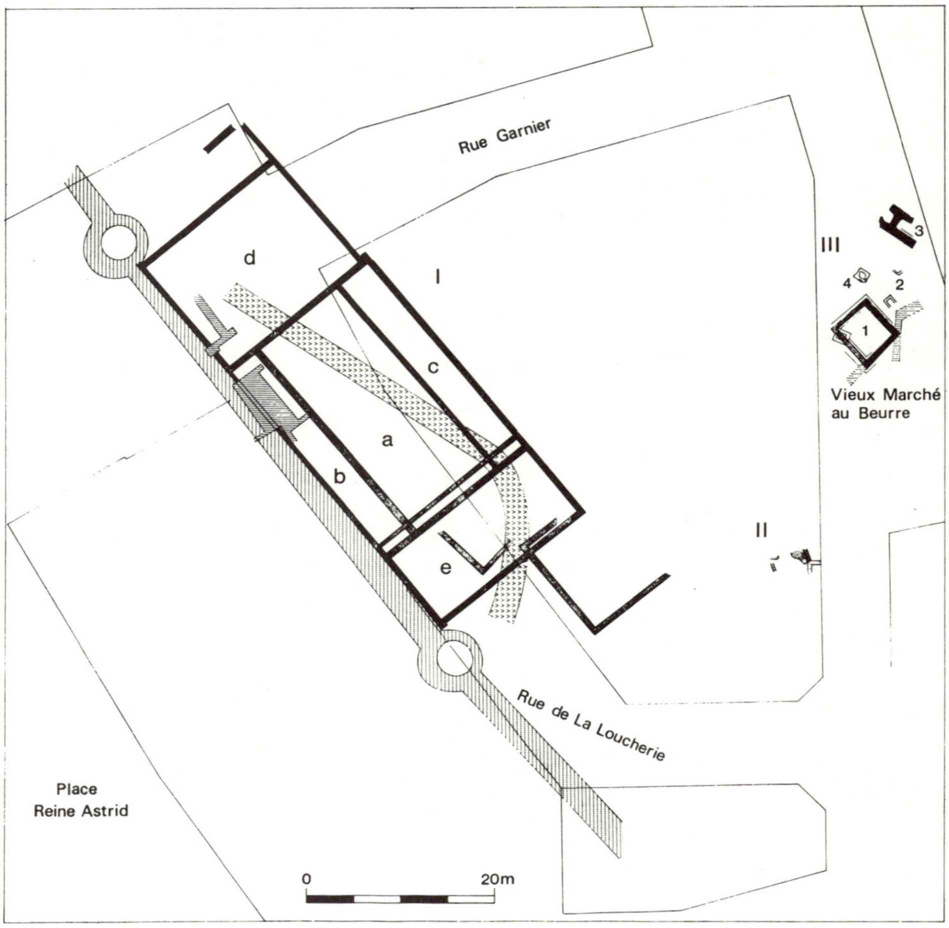

Fig 62 Tournai: the late Roman wall-circuit and the buildings at La Loucherie (after Mertens & Remy 1974, fig 14)

Fig 63 Tournai: the outer wall-facing of a tower of the late Roman wall-circuit (after Mertens & Remy 1974, fig 10)

blocks of stone 0·25–0·30 by 0·12–0·15 m set in yellow mortar. The foundation, of stone blocks and tiles, set in pink mortar, was laid on a drystone bed of blocks of limestone schist. The fragmentary state of the wall did not enable any repairs to be identified. The date suggested for it is the end of the 3rd century—the second quarter of the 4th century.

Rheims

The rescue excavations carried out in 1972–6 at Rheims located a portion of the late Roman walls near the southern monumental arch, the Porte Bazée, from an earlier period. Neiss (1976) presented the results of this work at the 101st Congrès National des Sociétés Savantes and the following details come from his paper.[3] The total length of the defences, calculated from their probable course, was about 3000 m (Fig 53). The wall, with a maximum width as far as could be established of 2·50 m, rested on a foundation consisting of a mass of unshaped stone and various construction materials, including fragments of tile (Fig 64). On this foundation was a course of reused blocks, about 0·25 m thick. Above this, five courses of squared blocks formed the lower part of the wall, making a total height of 3 m. Most of the blocks came from funerary monuments: they were roughly shaped to form courses 0·60 m high, and were set together without mortar. On top of this base the wall proper was built in the usual way and a bank of material behind was heightened in step with the stages of its construction. The facings were of small regular blocks of stone measuring 100 × 180/250 mm, and the

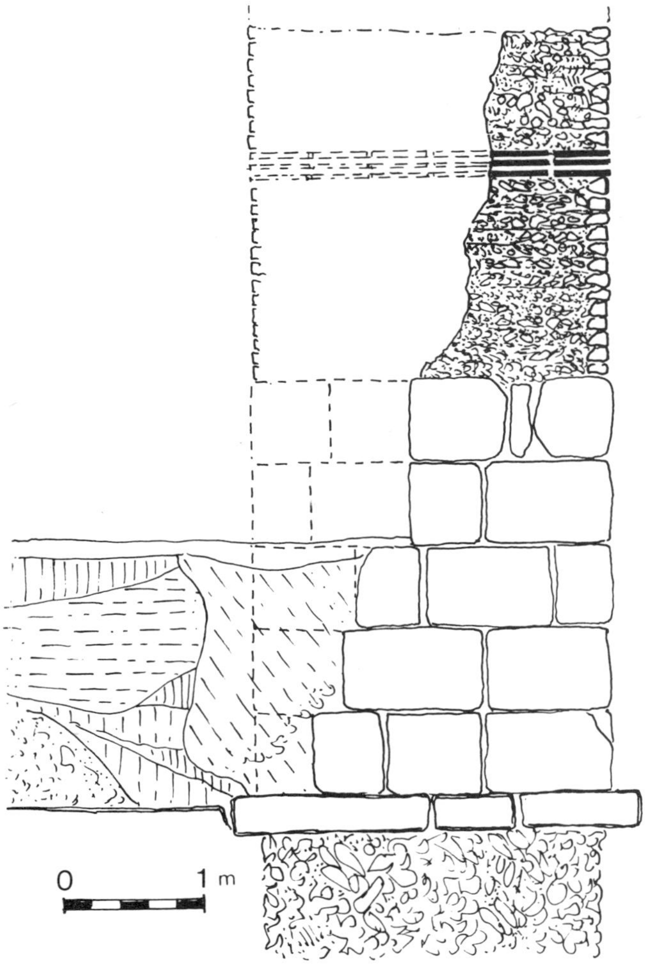

Fig 64 Rheims: section of the late Roman walls (after Neiss 1976)

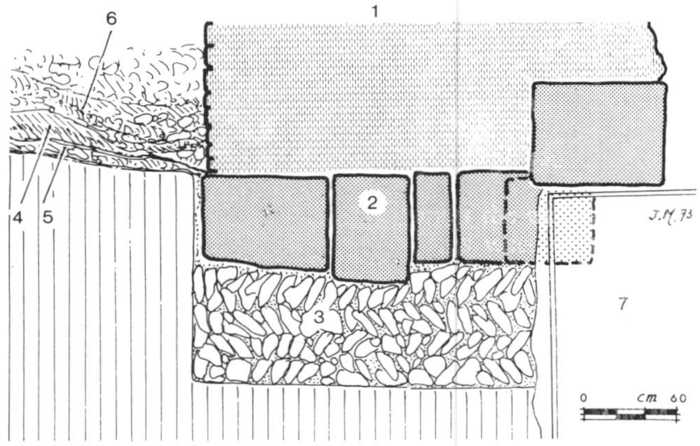

Fig 66 Arlon: section of the late Roman wall: 1 rampart core; 2 course of sculpted blocks; 3 foundation

core was composed of stones and mortar. At every 1·50 m were courses of flat bricks (containing small holes), measuring 400 mm long, 300 mm wide, and 50 mm thick. They were always set so that the shorter side was exposed at the face. The thickness of the mortar layer which bonded the tiles together was 30–40 mm, though only 20 mm in the joints between the stone blocks of the facings. It is not known how high the wall was originally, nor whether there was a defensive ditch. Archaeological discoveries have not provided a precise construction date for the defences.

Arlon

The late Roman wall-circuit at Arlon is famous for the number of Roman sculpted blocks it has produced, and in recent excavations it has lived up to its reputation. The excavations also established several interesting technical details: the wall was 3·80 m thick, and had many towers which were spaced along it at intervals of approximately 30 m. The diameter of the towers at foundation level was 9·50 m, though the superstructure was only 8·50 m in diameter. Since this wall-circuit with its towers was situated on a hill-slope the exterior surfaces have usually been destroyed, and consequently the external facing has never been found. The internal facing, on the other hand, is well preserved and comprises carefully-dressed small blocks of stone (Fig 65). The oblique marks of the mason's chisel

Fig 65 Arlon: interior facing of one of the towers of the wall-circuit

are still sharp. No tile-courses have been found. The foundation, 0·70 m deep, was of regular courses of stone blocks laid slantwise, capped by a course of large stones or reused blocks (Fig 66). No trace of rebuilding has so far been revealed, and precise evidence of dating has not been forthcoming (Mertens 1973, 8–14).

These few examples from the northern part of Gallia Belgica, though they may add very little that is new, do nonetheless allow some interesting conclusions to be drawn. Firstly, the wall-circuits were all built with a great deal of care, with small regular blocks of stone set in strong mortar. They do not therefore appear to have been hastily built, as the reuse of material from dismantled earlier monuments might lead one to suppose. Secondly, they have a great number of towers and, apart from the exceptions due to local technical peculiarities—for example at Bavai or Famars—they are all circular. Thirdly, the builders took account of pre-existing monuments, by including them and using them within the fortifications. Sometimes the walls were built in a single campaign—Tongres and Arlon—while sometimes they show traces of successive reconstruction and addition—Bavai and Famars—illustrating their adaptation to new demands.

The dating of these adaptations remains the great problem. The construction technique seems to be too heterogeneous to allow for precise dating. At Tongres, one of the beams from a tower foundation has been analysed by radiocarbon: the date it produced was 1700±50 BP which indicates, after calibration, a date around AD 260. For the walls of Bavai and Famars, where it has been possible to establish a relative chronology, the proposed date for the first phase is the last quarter of the 3rd century. The second phase is normally considered to belong to the Constantinian period, and the third may reflect rebuilding and restoration carried out in the second half of the 4th century. These dates highlight the most important moments in the evolution of the defensive system in the whole of northern Gaul (Brulet 1978, 3–8). One should note that in Germania Secunda the walls of Maastricht have been dated by Bogaers to after AD 355.[4]

To round off this brief study of urban defences in Gallia Belgica, I would like to make two further points. In the first place, romanization only spread through the north of Gaul very slowly. Nearly a century had passed by after the initial conquest before the development of communities which resembled towns, and it was well into the 2nd century before one or two of these new towns received a wall-circuit, a monument which was primarily purely symbolic and decorative. Gallia Belgica lay well inside the frontier, and thus benefited from the *pax Romana.* Then, after the invasions of the second half of the 3rd century, the position was completely reversed. Northern Gaul became a frontier zone, the heart of the empire's defences. Roads and countryside were dotted with forts; towns, villages, even hamlets, withdrew behind new walls—reduced in scale, but effective. These walls had their role to play for many centuries to come, into the medieval period.

Notes

1 Concerning the origin and evolution of these gates, comparison with Italian reliefs is very interesting (Rebecchi 1978, 125–44). I would like to thank Mr Rebecchi for having obtained for me a copy of the Caserta Vecchia relief.

2 The unfinished appearance of the architectural ornamentation could be a sign of the interruption of work in the aftermath of the troubles at the end of the 2nd century (Meyer-Plath in Gose *et al* 1969, 86).

3 Neiss 1976. I would like to thank the author for having sent me a duplicated copy of this paper.

4 Bogaers 1959, 59–60; the same author is less precise—'at the earliest at the beginning of the 4th century'—in Bogaers & Rüger 1974, 186.

References

Amand, M, 1951 Turnacum Romanum, *Antiq Class,* **20,** 383–408

———, 1974 L'approvisionnement en eau de Tournai romain, *Archaeol Belgica,* **143,** 5–47

———, 1977 Urban sites of Roman origin in Belgium, in *European towns* (ed M W Barley), 159–65

Bernard, H, 1975 Thérouanne, ville morte, *Archéologia,* **81,** 41–9

Bersu G, & Unverzagt, W, 1961 Le castellum de Fanum Martis, *Gallia,* **19,** 159–90

Biévelet, H, 1964 L'exploration archéologique de Bavai, *Revue du Nord,* **46,** 193–205

Blanchet, A, 1907 *Les enceintes romaines de la Gaule,* Paris (repr 1979)

Bogaers, J, 1959 in *Nieuwsbull Kon Nederlandse Oudheidk Bond,* 59–60, 85–6

Bogaers, J, & Rüger, C B (eds), 1974 *Der Niedergermanische Limes. Materialien zu seiner Geschichte,* Bonn

Brulet, R, 1975 *Recherches archéologiques sur le Bas-Empire romain dans les civitates Turnacensium, Camaracensium, et Tungrorum,* unpubl doctoral thesis, Univ Louvain

———, 1978 Fortifications et défense du territoire au Bas-Empire en Gaule septentrionale, *Caesarodunum,* **28,** 3–20

Butler, R M, 1959 Late Roman town walls in Gaul, *Archaeol J,* **116,** 25–50

Clavel, M, & Lévèque, P, 1971 *Villes et structures urbaines dans l'occident romain,* Paris

Crema, L, 1959 L'architettura romana, in *Enciclopedia classica,* **12,** 1, Turin

Cumont, F, 1914–19 Comment la Belgique fut romanisée, *Annales Soc Archéol Bruxelles,* **28,** 81–96

Cüppers, H, 1969 Die Trierer Römerbrücken, *Trierer Grabungen und Forschungen,* **5**

———, 1973 Die Stadtmauer des römischen Trier und das Graberfeld an der Porta Nigra, *Trierer Zeitschr,* **36,** 133–222

———, 1977 Die römische Stadtmauer, *Führer zu vor- und frühgeschichtlichen Denkmälern,* **32,** *Trier,* 74–9

Delmaire R, 1976 Etude archéologique de la partie orientale de la cité des Morins, *Mémoires Comm Pas-de-Calais,* **16,** 133–9

de Schaetzen, P, 1955 De datering van de eerste Romeinse stenen omheining van Tongeren, *Oude Land van Loon,* **10,** 1–9

Gose, E, Meyer-Plath, B, Steinhausen, J, & Zahn, E, 1969 Die Porta Nigra, *Trierer Grabungen und Forschungen,* **4**

Gosselin, J Y, Seillier, Cl, & Leclercq, P, 1976 Boulogne antique. Essai de topographie urbaine, *Septentrion,* **6,** 5–15

Harmand, L, 1961 Le problème du rempart urbain dans les provinces occidentales de l'Empire aux deux premiers siècles, *Atti 8 Congresso Archeol Class,* Rome, **3,** 195–202

Hassel, F J, & Künzle, E, 1975 Zur Karte der Städte besonderer Rechtsstellung im Westen des römischen Reiches, *Ausgrabungen in Deutschland,* **3,** 339–42

Hollande, M, 1976 *Essai sur la topographie de Reims*

Johnson, J S, 1973 A group of late Roman city walls in Gallia Belgica, *Britannia,* **4,** 210–23

Lugli, G, 1960 La Porta Nigra di Treviri, *Riv Instituto Archeol e Storia d'Arte,* **18,** 97–101

Mertens, J, 1964 Enkele beschouwingen over Limburg in de Romeinse tijd, *Archaeol Belgica,* **75,** 17–24

———, 1967 Een Romeins tempelcomplex te Tongeren, *Kölner Jahrb für Vor-und Frühgeschichte,* **9,** 101–6

———, 1973 Le rempart romain d'Arlon, *Speculum,* **7**

———, 1977 De laat-romeinse stadtsmuur van Tongeren, *Archaeol Belgica,* **196,** 49–54

———, 1980 Recherches récentes sur le Limes en Gaule Belgique, in *Roman frontier studies 1979* (eds W S Hanson & L J F Keppie), Brit Archaeol Rep, **S71,** 2, 423–70

Mertens, J, & Remy, H, 1974 Tournai, fouilles à la Loucherie, *Archaeol Belgica,* **165,** 22–3

Neiss, R, 1976 *L'enceinte du Bas-Empire à Reims: fouilles actuelles,* paper presented to the 101st Congrès National des Sociétés Savantes, Lille

Raepsaet-Charlier, M Th & G, 1975 Gallia Belgica et Germania Inferior. Vingt-cinq années de recherches historiques et archéologiques, in *Aufstieg und Niedergang der römischen Welt* (eds H Temporini & W Haase), **2,** 4, 3–299

Rebecchi, F, 1978 Précédents italiques de la Porta Nigra de Trèves, *Caesarodunum,* **28,** 125–44

Stillwell, R, MacDonald, W L, & McAllister, M H, 1976 *The Princeton encyclopedia of classical sites*

Ternes, C-M, 1978 Les origines de la ville de Trèves: le problème des relations entre les oppida préromains et l'urbanisme en plaine, *Caesarodunum,* **28,** 61–8

Ulrix, F, 1962/63 Comparaison des plans des villes romaines de Cologne, Trèves et Tongres, *Kölner Jahrb für Vor- und Frühgeschichte,* **6,** 58–70

Vanvinckenroye, W, 1964 Opgravingen te Tongeren 1963–1964, *Publicaties Prov Gallo-Romeins Museum Tongeren*, **8**,
———, 1975 Tongeren, Romeinse stad, *ibid*, **23**
Vasselle, F, 1962 L'enceinte urbaine du Bas-Empire de Samarobriva (Amiens), *Celticum*, **6**, 323–42
Vasselle, F, & Will, E, 1960 L'urbanisme romain à Samarobriva-Amiens, *Revue du Nord*, **42**, 337–52
Wightman, E, 1970 *Roman Trier and the Treveri*
———, 1976 Der Meilenstein von Buzenol, eine Inschrift aus Mainz und die Rechtsstelle des römischen Trier, *Trierer Zeitschr*, **39**, 61–8
Will, E, 1960 Les remparts romains de Boulogne-sur-Mer, *Revue du Nord*, **42**, 363–79
———, 1961 Informations archéologiques. Circonscription de Lille, *Gallia*, **19**, 274–87
———, 1962a Recherches sur le développement urbain sous l'Empire romain dans le nord de la France, *ibid*, **20**, 79–101
———, 1962b Les enceintes du Bas-Empire à Bavay, *Revue du Nord*, **44**, 391–401
———, 1963 Les enceintes du Bas-Empire à Bavay, *V Congressus Internat Limiti Romani, Zagreb*, 99–109
Zahn, E, 1977 Porta Nigra und Simeonsstift, *Führer zu vor- und frühgeschichtlichen Denkmälern*, **32**, *Trier*, 61–74

The Aurelianic wall of Rome and its analogues

Malcolm Todd

In AD 270 the Danube frontier was twice broken by barbarian raiders. On the second occasion the invaders reached Italy and began to threaten the cities of the north. At the beginning of the year Lucius Domitius Aurelianus had been proclaimed emperor after the death of Claudius II in an outbreak of plague. The threat to Italy was quickly dealt with and the barbarians were dispersed. But the threat to the unprotected heart of the Empire remained and it was increasingly clear that the safety of Rome itself could no longer be guaranteed in the changed world of the later 3rd century. The *urbs* did not have the protection of walls, having centuries before outgrown the defences of Republican Rome. In 271 work began on the construction of a great circuit of walls which was to surround all of the city east of the Tiber and part of the Janiculum to the west. These walls were to be a major determinant in Roman topography down to the Baroque and are still one of the most impressive monuments of the city (Richmond 1930).

Aurelian's great wall can still astound by its sheer scale (Fig 67). It is eighteen kilometres (twelve miles) long and in its original form was 6·5 m high and 3·6 m thick. It was built of brick- or tile-faced concrete, a considerable proportion of the tiles being derived from earlier structures. It possessed a parapet and merlons, short stretches of which can still be seen in the later brickwork. In certain sectors the wall contained a barrel-vaulted gallery above a height of 3 m, immediately above which lay the rampart-walk. The wall was apparently built in sections measuring 4·5 to 6 m long and 1·3 to 1·8 m high. The joins between these work-sectors were normally carefully concealed at the facing and are now usually difficult to identify. In general the standards of workmanship evident in the wall

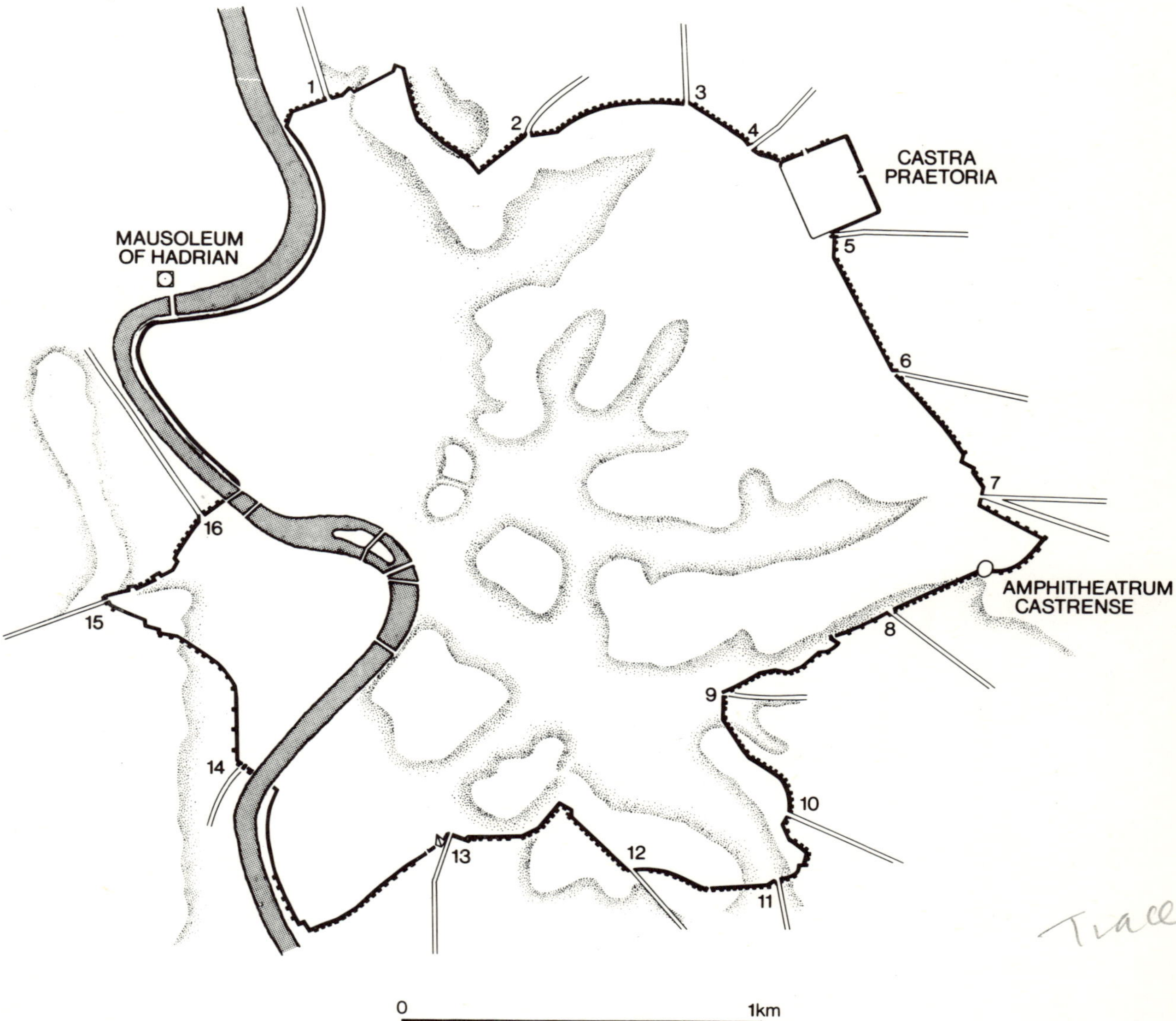

Fig 67 The defences of Rome: 1 Porta Flaminia; 2 P Pinciana; 3 P Salaria; 4 P Nomentana; 5 P Chiusa; 6 P Tiburtina; 7 P Praenestina; 8 P Asinaria; 9 P Metrobia; 10 P Latina; 11 P Appia; 12 P Ardeatina; 13 P Ostiensis East; 14 P Portuensis; 15 P Aurelia; 16 P Septimiana

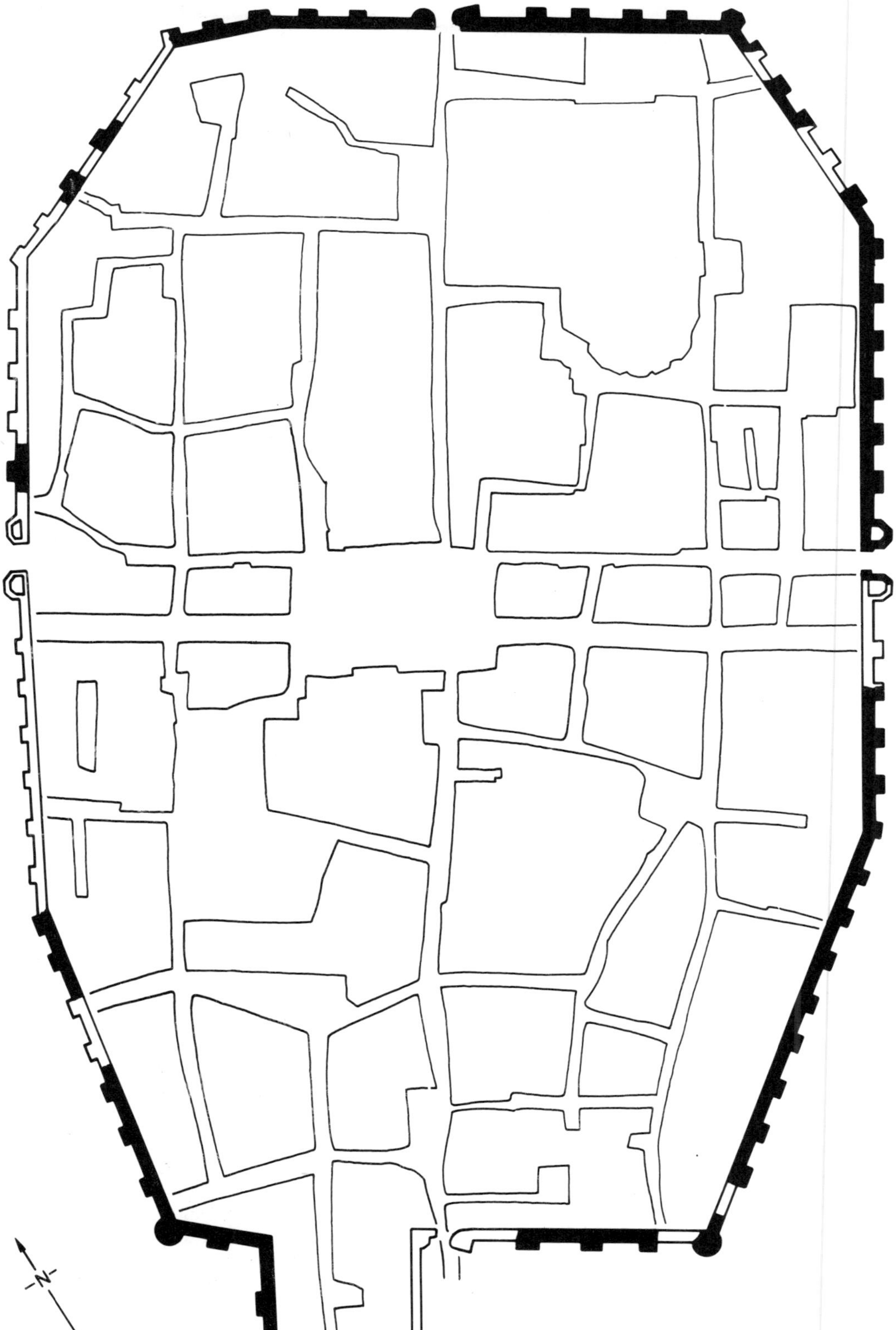

Fig 68 Barcelona: the late Roman defences

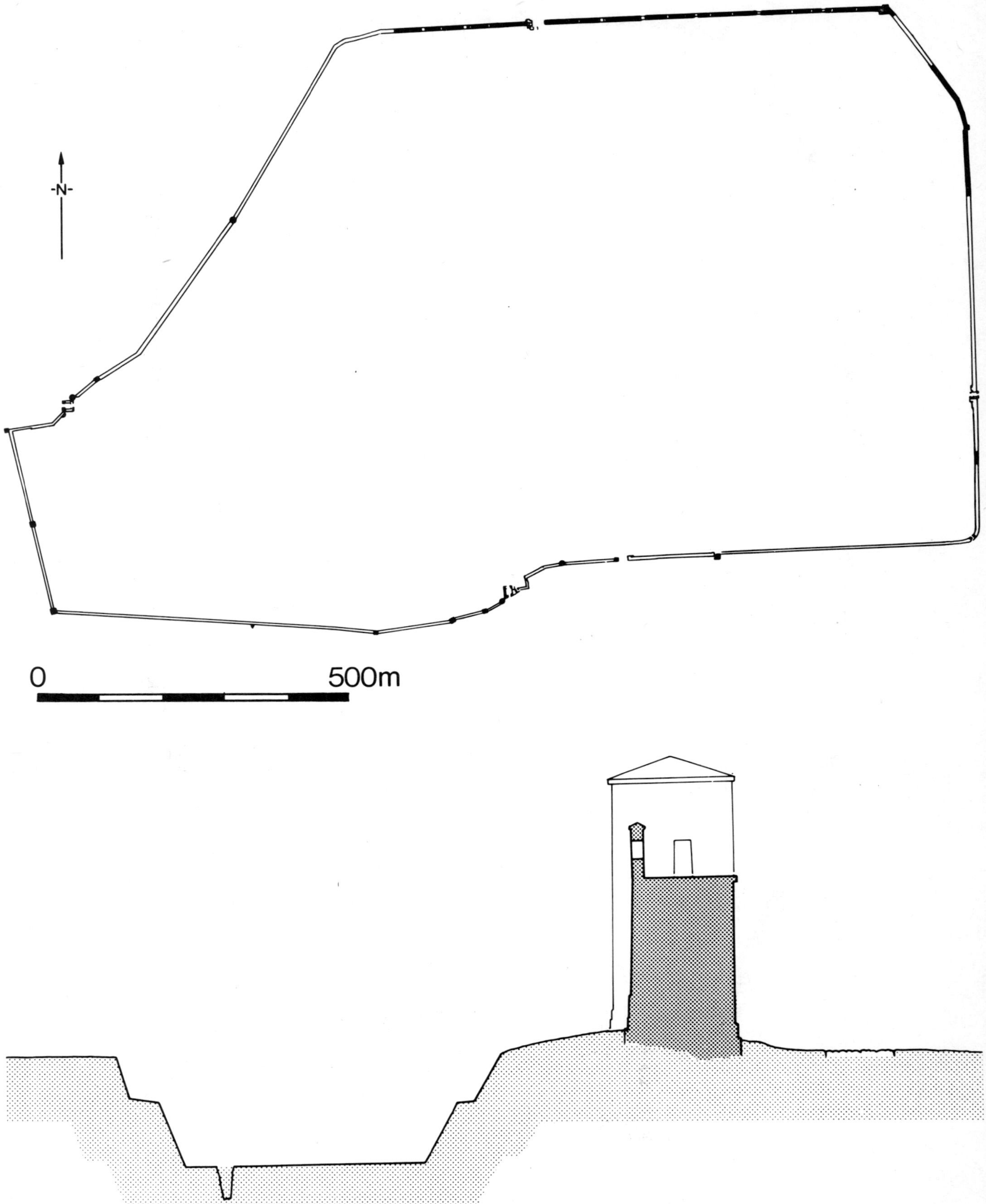

Fig 69 The defences of Paestum

Fig 70 The stone relief from Pinara (Lycia)

are high. The concrete core of the wall was tightly packed in behind the tile-facing, and horizontal layers of tiles running through the thickness of the wall gave a stability which has defied all but the most severe earthquakes. There is no clear evidence of an earth rampart to the rear. In certain sectors, for instance south of the Porta Chiusa, the wall appears to revet a mass of earth. These are best seen, however, as natural banks against which the wall has been set or into which its rear face has been inserted.

An enormous number of towers were included in the original building scheme. The total number is likely to have been 381. They were evenly spaced at an average interval of about 30 m along the entire circuit except on the Tiber front. The majority followed a standard pattern, measuring 7·6 m wide and projecting 3·35 m from the wall-face. The lower part of each tower was a mass of concrete solid to the level of the rampart-walk. Over this lay a chamber which extended to the rear face of the wall and thus measured a little under 7×7·6 m. The chamber was usually lit by two round-headed windows in the front, one at the rear, and one in each side. The towers on the galleried wall were of the same dimensions and had a similar outward appearance, though their internal planning was perforce slightly different. In none of the towers is there any evidence of a means of direct access from the ground to the chamber above. At the gates alone was it possible to reach the rampart-walk.

The gates as originally planned, eighteen in all, followed a simple and long-established pattern, in which a single or double portal was flanked by a pair of semicircular fronted towers. Gates on the main roads, Portae Appia, Flaminia, Ostiensis East, and Portuensis, were given double portals, the rest single openings only. I have argued elsewhere (Todd 1978, 35, 72) that all the Aurelianic gates fall within a long tradition extending back to the later 1st century BC at least. In many respects the gates of Aurelian's wall are among its least remarkable features and there is nothing to be added to earlier discussions of them.

Major changes were made to Aurelian's wall less than half a century after its construction. A tall galleried wall was added to the original curtain, increasing its height to about 15 m and, in places, to rather more. The towers, too, were heightened, though no uniform plan was followed. Certain gates, notably Porta Appia and Porta Asinaria, received special attention in this rebuilding in the form of monumental towers four storeys high (24 and 18·5 m respectively). Since Richmond's brilliant account of the Roman defences, it has been generally accepted that these changes were wrought by Maxentius during his six year tenure of Rome, from 306 until his defeat at the Milvian Bridge in 312. I would now be more inclined to argue that Constantine himself may have been largely responsible for this major programme of work. Styles of brickwork alone do not allow us to decide upon a date before or after 312 and little other evidence is available to us. I am increasingly doubtful, too, as to whether Maxentius would have had the means to complete the work in only six years. Some of the elements in the scheme, at least, may well be Constantinian.

There is little uncertainty about the date of the final major embellishment of the wall. In the early years of the 5th century, the double-portalled gates were reduced to single entrances by the insertion of stone gate-houses and the lowest storeys of their towers were handsomely encased in stone or, in the case of Porta Appia, in white marble. Grandiloquent inscriptions spanned at least three of the gates: Portae Tiburtina, Praenestina, and Portuensis (*CIL*, **6**, 1188–90). These must have been inserted in 401 or early in 402. Much of this work, if not all of it, was probably completed before Honorius's triumphal entry on his first visit to Rome in 404.

The place of Aurelian's wall in urban fortification

The late 3rd century defences of Rome present a number of problems, with only one of which this paper is concerned. In several major respects Aurelian's wall stands outside the main tradition of city walls in the west. The very size of the *urbs* posed problems for the builders which did not have to be faced elsewhere. The solid rectangular tower type chosen for the enceinte is found only occasionally in the western provinces and the close spacing of the towers, though it is matched in one group of 3rd century walls, is not a common western characteristic. The use of a gallery in certain stretches of the original Aurelianic structure is reminiscent of Hellenistic practice in the Aegean and in Asia Minor, but it finds no analogy among the surviving western walls of the late Empire. Several of the general characteristics of the Roman wall are evident in certain circuits in northern Spain, notably Barcelona (Fig 68), Lugo, Astorge, Saragossa, and León. But there are significant points at which the comparison breaks down, particularly on the details of the tower type. Unfortunately we still know remarkably little of the history of urban fortification in Hellenistic and Roman Italy. That the Aurelianic builders may have found inspiration for their

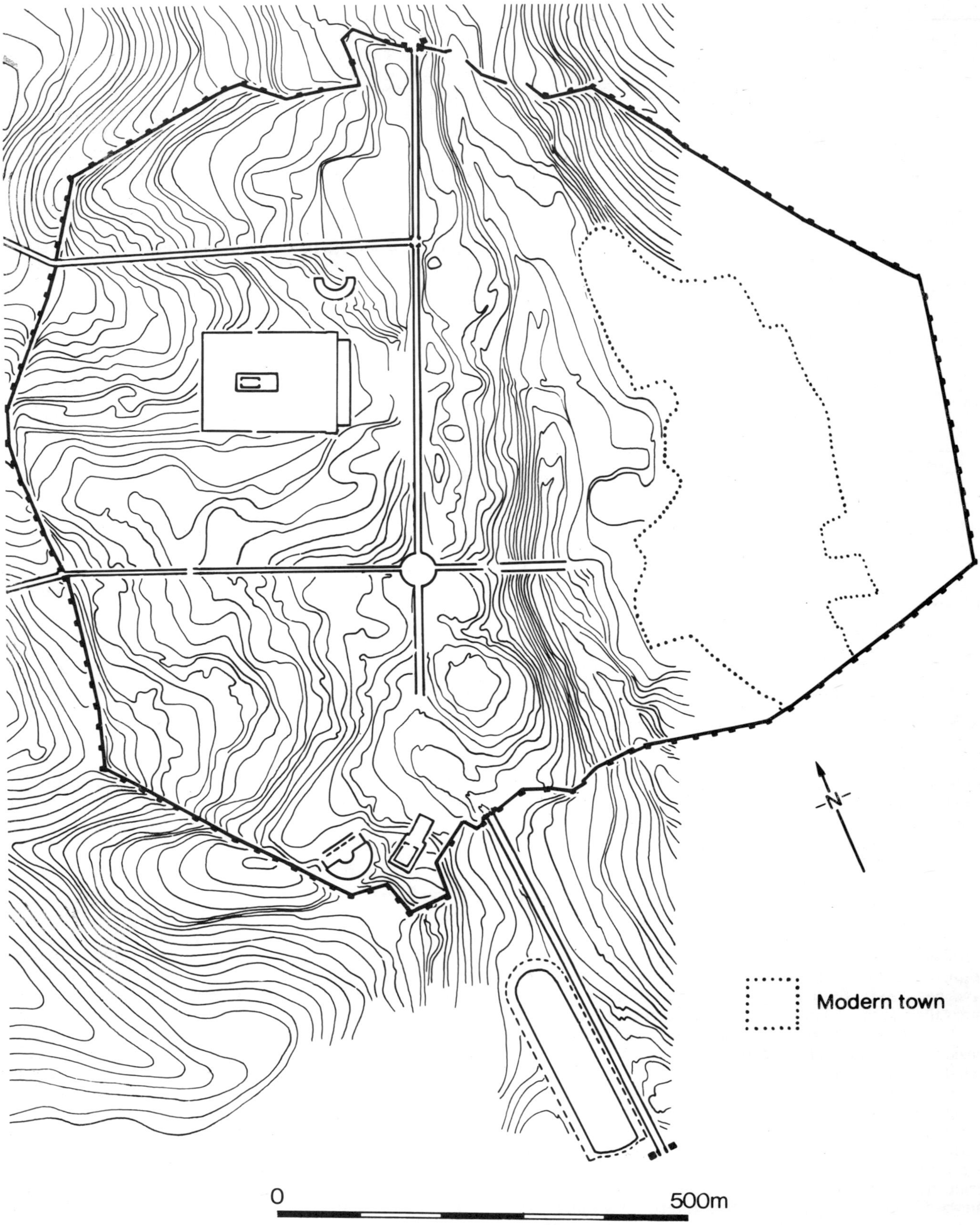

Fig 71 The defences of Gerasa

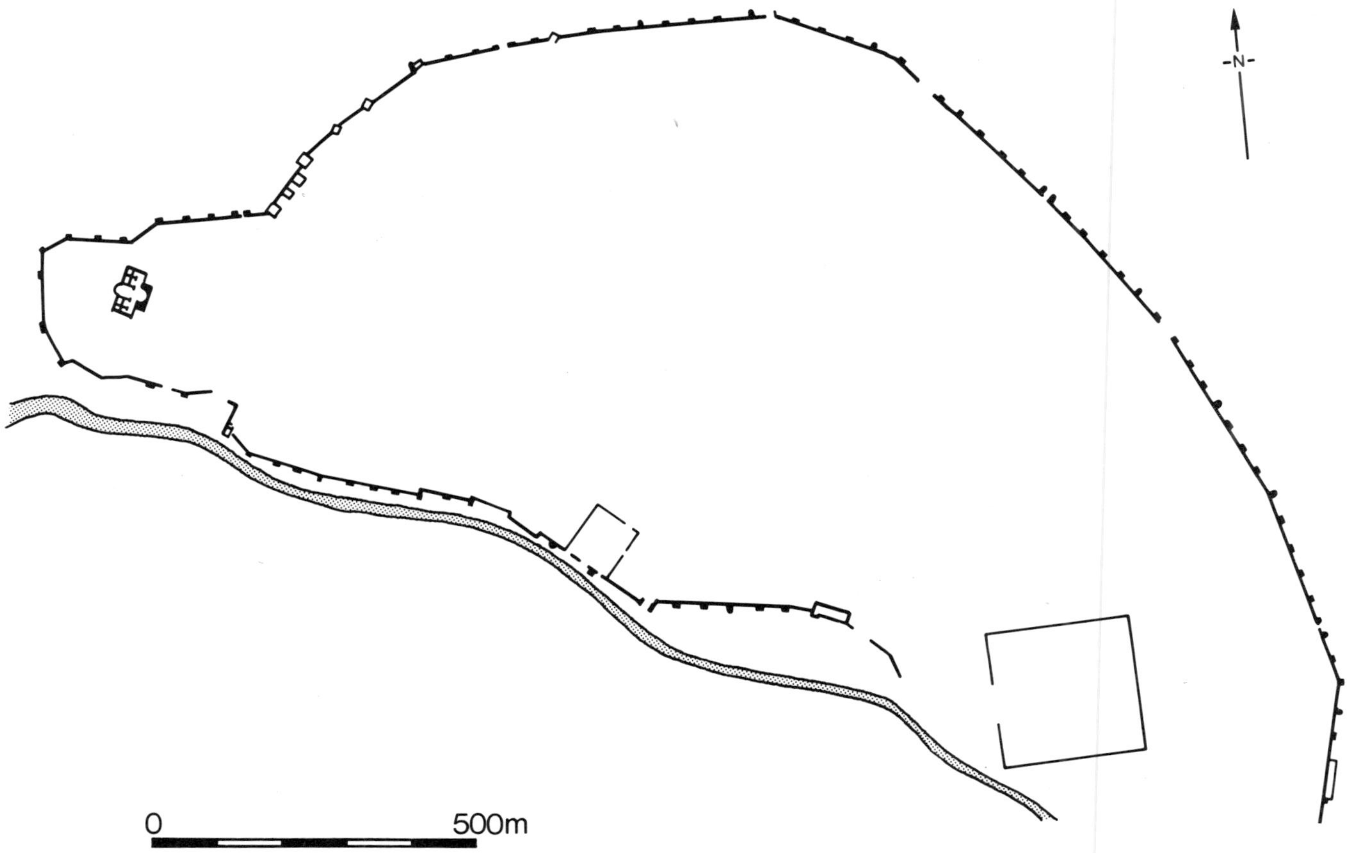

Fig 72 The 3rd century defences of Palmyra

defences in Hellenistic fortification in South Italy (eg Paestum, Fig 69) and Sicily, for example, seems a not unreasonable hypothesis *a priori.* But those fortifications were themselves a reflection of a very old defensive tradition and to study it we must turn elsewhere.

So many well preserved urban circuits exist in western Europe that we are liable to forget that there existed a vigorous and distinctive tradition of defensive architecture in the eastern provinces, especially in those which were in close contact with the Parthian world (Fig 70). If we move eastwards to Asia Minor, to the Euphrates frontier and still further beyond the frontiers of the Empire, a combination of all the elements we have observed at Rome is apparent in a wide range of cities, forts, citadels, and sacred places. Settlements of the Hellenistic period in the east have been little studied. Still less have their defences been subjected to detailed survey and analysis—most unfortunately, as the cities in the Euphrates valley in particular lie at a most fruitful meeting point of east and west. Enough is known of their defences, however, to make it plain that here are present all the main features of Aurelian's wall, especially the rectangular towers, often solid in their lower storeys and almost invariably set close together. Several of the cities in Asia Minor and the great plain of Iraq were large and sprawling. Their defensive circuits were consequently long, much longer than was normal in the Roman west. Warfare commonly involved protracted sieges. Sometimes, indeed, it consisted of little else. The crucial operations of several Imperial campaigns in the east were sieges. The two failures of Septimius Severus at Hatra (Dio lxxvi, 10–12) and the Persian siege of Amida so vividly described by Ammianus (xix, 2–8) are dramatic examples, but there were many others.

Systems of close-set rectangular towers are so frequently found girding the cities of Greece, Asia Minor, and the lands eastward to the valleys of Tigris and Euphrates that we may here confine our attention to a few striking instances. Gerasa in the Decapolis (Fig 71) possessed a fine circuit of walls 3·6 km long, probably constructed in the second half of the 1st century AD and surviving to provide the basis for the fortifications of the late 4th century city (Kraeling 1938; Bietenhard 1977). The fine and complicated circuit which surrounds the city of Palmyra in the Syrian desert (Fig 72) is beset by many problems which can only be resolved by further excavation. But the walls of the 6th century AD are probably a rebuild of defences which were constructed during the spectacular rise of Palmyra in the period of the early Empire (Wiegand 1932; Michalowski 1970; Richmond 1963). Dura, on its commanding position above the Euphrates (Fig 73), was well defended by nature as well as by a circuit of walls and towers of Hellenistic type. The towered 3·3 km circuit was first constructed at the foundation of the Macedonian city by Seleucos Nikator about 280 BC, but the subsequent vicissitudes of Dura (in Parthian hands from 114 BC and recovered by Rome in AD 165) will have left their mark on the defences, even though the detailed building history has still to be worked out. The walls of Ephesos (Miltner 1958) may represent the myriad cities of the Aegean and Asia Minor (Fig 74). These

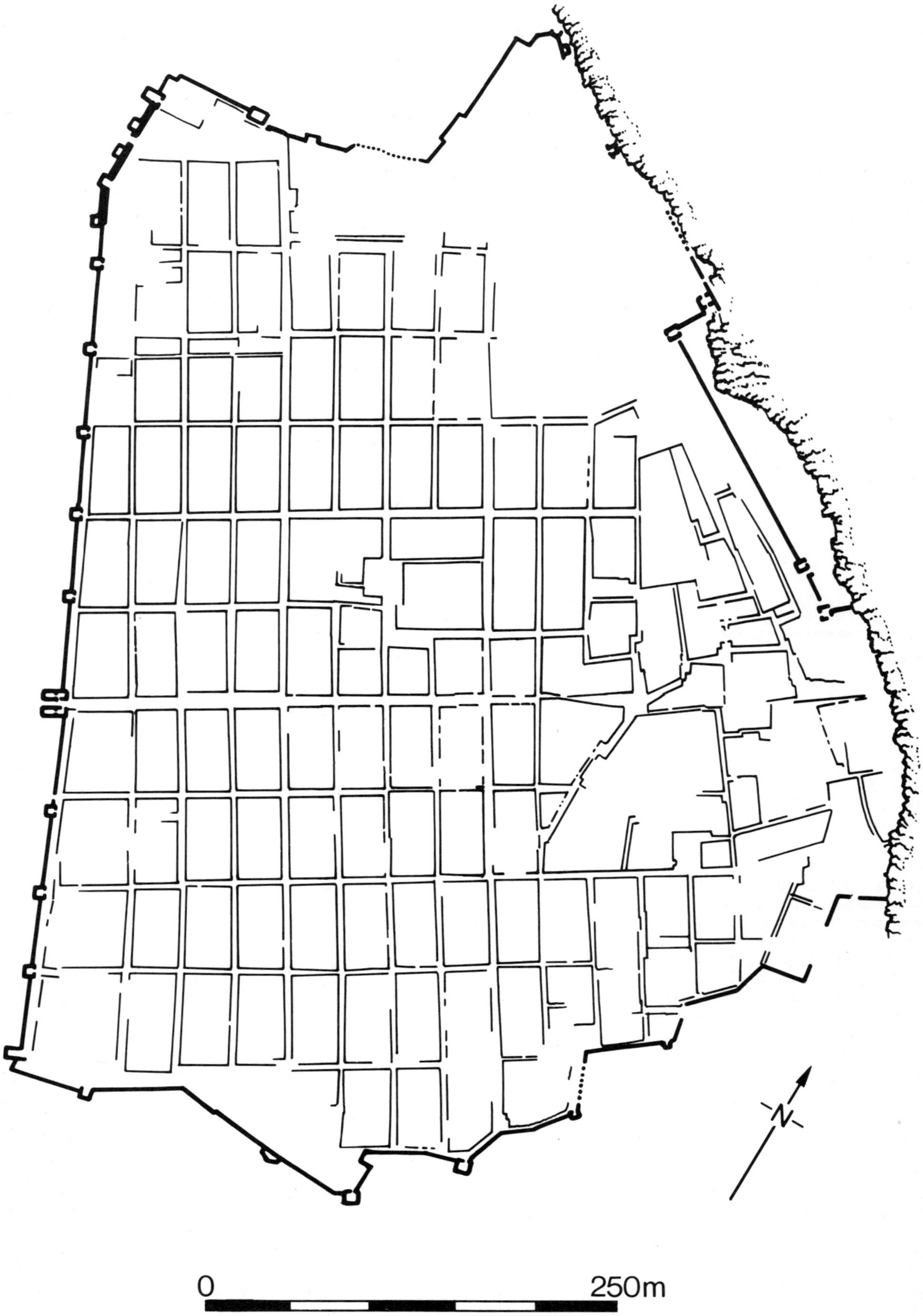

Fig 73 The defences of Dura-Europos

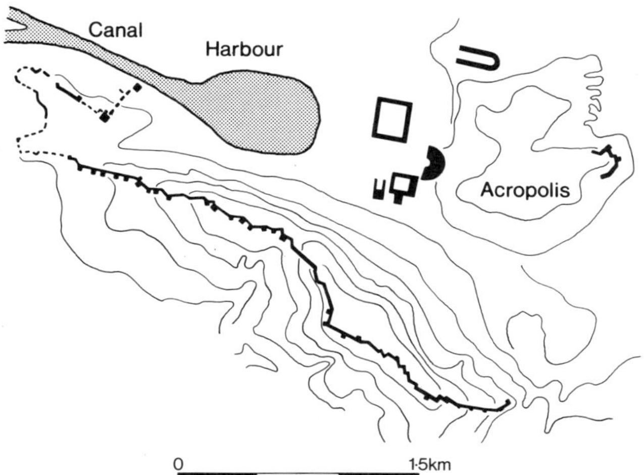

Fig 74 The defences of Ephesos

defences belong to the great replanning of the city about 295 BC by Lysimachus and apparently continued to surround the opulent maritime city of the earlier Roman Empire. Towered circuits of the same type are evident all over Asia Minor, for example at Perge, Side, Miletus, Priene, and Aphrodisias.

The history of all these circuits is known only in the barest outline. There are, however sufficient indications that the provision of regular systems of projecting rectangular towers, often solid in their lower storeys, extends back to at least the Seleucid era and, in the Persian territories, to the Achaemenid period. The origins of this type of fortification lie much further back in remote antiquity. Superficially at least, the main elements in the scheme were already present, for instance, in Troy VI and Kouklia (Old Paphos). But there is no doubt that they passed into wider currency in the period which opened with the supremacy of Macedon. One of the finest surviving works of that period is the circuit of walls at Demetrias in Thessaly, founded by Demetrius Poliorcetes in 293 BC (Fig 75). As will emerge in the rest of this paper, it is entirely appropriate that a work of the great besieger himself should represent the type.

The dating and building sequence still require refinement, but there is no doubt that in the first three centuries of the Roman Empire circuits of this type formed the commonest model of defence provided for the large urban centres of the Greek east. They were, however, by no means confined to cities. They were also constructed around citadels and religious sites in both the Hellenistic and the Parthian worlds. Roman forts on the eastern frontiers and on the roads leading to them also reveal many examples (for instance, Ain Sinu; *Iraq*, **21** (1959), 214; Fig 76).

In their outward form, the defences of large cities like Gerasa, Dura, Palmyra, and Ephesos provide the closest analogies for the 3rd century defences of Rome. The walls of the eastern cities were designed above all to enable large communities to survive long sieges. The strong construction and the close arrangement of their towers were particularly well suited to the needs of artillery, as well as the requirements of archers and spearmen. It may be argued that the defences of Rome owed nothing to this Helleno-Parthian tradition of fortification, that they were inspired by the Hellenistic circuits of southern Italy, for example, or that they were no more than a variant of western fortifications adapted to the peculiar circumstances of the *urbs*. We are, however, not compelled to depend on the crudities of purely typological argument in suggesting a link between Aurelian's wall and the easternmost fringes of the Roman world.

In the same year as work began on the Roman defences, Aurelian mounted his expedition against Zenobia and Vaballathus. His advance on Palmyra went through the Balkans, on to Byzantium, Ancyra, and Antioch, and was interrupted only by a brief siege at Tyana. After a successful engagement near Antioch, Aurelian continued his march to Emesa, where again he defeated the Palmyrene forces. Zenobia withdrew to Palmyra itself and made preparations for a siege. But she had overestimated the loyalty of the Palmyrenes and her Persian allies. The gates of Palmyra were soon opened to Aurelian, but not before he himself had been wounded in fighting outside the walls. In the following year, 272, Palmyra again revolted against Rome and this time the city was given over to pillage and its powerful walls were destroyed.

Aurelian himself, then, had first-hand knowledge of the eastern urban enceintes. He had seen their effectiveness in siege-warfare. Neither Tyana nor Palmyra had fallen to the assault of the besiegers. At the former the gates were opened by a traitor, at the latter by a peace party. No doubt the strength of the frontier cities had impressed the commanders of earlier expeditions, notably Trajan and Septimius Severus. But Aurelian alone of those who campaigned in the east was also to be responsible for building a great urban circuit in the west.

The military campaigns of 271–2 may provide a plausible and immediate link between the eastern cities and the *urbs*. But there were in the later 3rd century links of a more general and thus more powerful kind between Rome and her Parthian frontier. As so many studies of the past thirty years have emphasized, the meeting ground of Rome and Parthia was one of the most fruitful creative centres of the ancient world. Ward-Perkins, himself responsible for several perceptive contributions to the subject, has put it thus (1965, 196):

> To the extent that we polarize the history of western Asia in Roman times into a confrontation of the two great powers, Rome and Parthia, we tend to obscure the part played by this cultural third force within the territories on either side of the political frontiers. These territories were indeed the middlemen for the passage of goods and ideas from west to east and from east to west. But they were also something very much more than that. They were a creative centre in their own right, a melting-pot and a forcing-house for many of the most vital of the new ideas that were to carry the ancient world forward into the Middle Ages.

Aurelian's great fortification may fitly take its place in the eastern heritage of the later Roman Empire, an early borrowing from that great complex of ideas, techniques, and materials which was to influence so deeply the art and architecture of eastern Rome.

Acknowledgements

I am much indebted to Professor T P Wiseman for reading a draft of this paper and to M J Rouillard for the drawings which illustrate the final version.

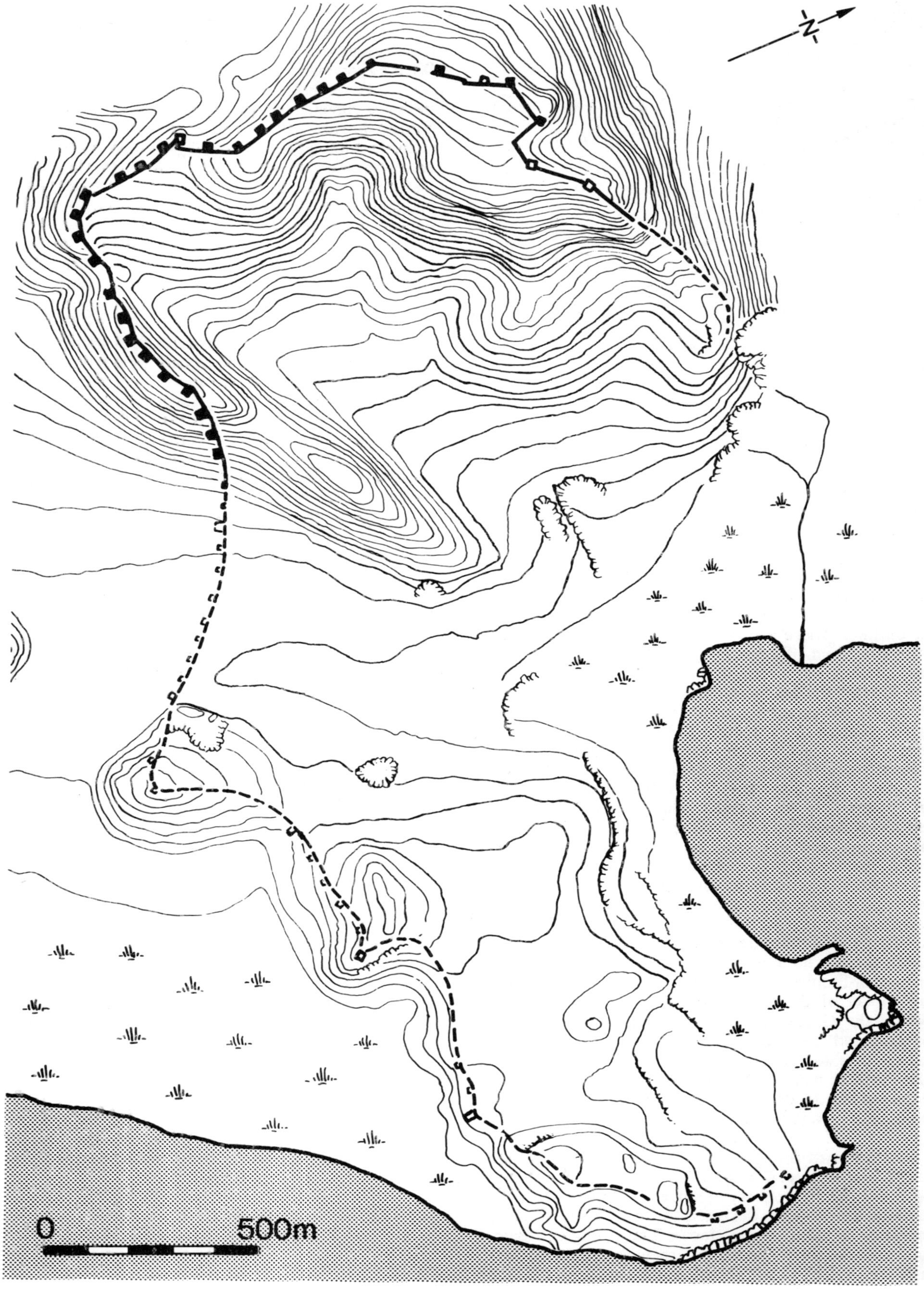

Fig 75 The defences of Demetrias (Thessaly)

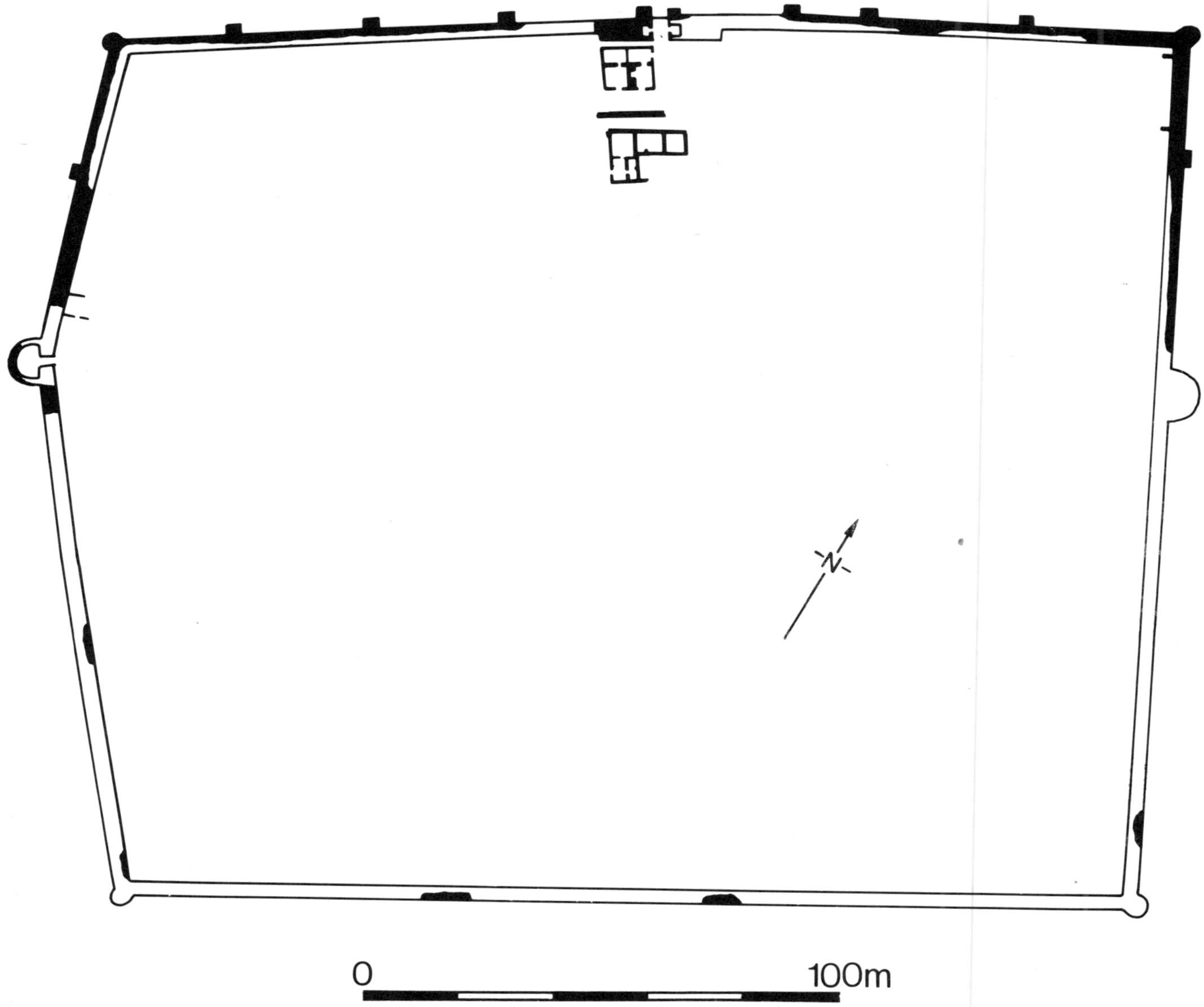

Fig 76 The castellum *at Ain Sinu (Iraq)*

References

Bietenhard, H, 1977 Die syrische Dekapolis von Pompeius bis Traian, in *Aufstieg und Niedergang der römischen Welt* (ed H Temporini), **2,** 8, 220–61

CIL *Corpus Inscriptionum Latinarum*

Kraeling, C H, 1938 *Gerasa: city of the Decapolis,* New Haven

Lawrence, A W, 1979 *Greek aims in fortification*

Michalowski, K, 1970 *Palmyra*

Miltner, F, 1958 *Ephesos,* Vienna

Richmond, I A, 1930 *The city walls of Imperial Rome*

———, 1963 Palmyra under the aegis of the Romans, *J Roman Stud,* **53,** 43–54

Todd, M, 1978 *The walls of Rome*

Ward-Perkins, J B, 1965 The Roman west and the Parthian east, *Proc Brit Acad,* **51,** 175–200

Wiegand, T, 1932 *Palmyra. Ergebnisse der Expeditionen von 1902 und 1917,* Berlin

Batavorum
Colonia Ulpia Traiana
Colonia Claudia Ara Agrippinensium
Traiectus
Atuatuca Tungrorum
Gesoriacum
Tarvenna
Turnacum
Antunnacum
Icorigium
Fanum Martis
Bagacum
Baudobriga
Nida
Beda
Moguntiacum
Samarobriva
Augusta Vermanduorum
Orolaunum
Noviomagus
Colonia Treverorum
Vicus Attiaensium
Rotomagus
Caesaromagus
Virton
Lopodunum
Augustodurum
Suessionum
Cratisium
Aliso
Castra Regina
Noviomagus Lexoviorum
Augustomagus
Durocortorum
Divodurum
Mediolanum Aulericorum
Lutetia Parisiorum
Ovilava
Tres Tabernae
Argentorate
Augusta Vindelicum
Condate Redonum
Agedincum
Suindinum
Cambodunum
Autessiodurum
Castellum Rauracense
Irgenhausen
Condevicnum
Caesarodunum
Avaricum
Augustodunum
Aventicum
Aguntum
Mediolanum Santonum
Augusta Praetoria
Verona
Vesunna
Cularo Gratianopolis
Augusta Taurinorum
Burdigala
Nemausus
Aquae Tarbellicae
Arelate
Aquae Sextiae
Tolosa
Forum Iulii
Iruna
Carcaso
Narbo Martius
0
500km

Late Roman urban defences in Europe

Stephen Johnson

This paper attempts only to make one or two main points about the construction of the large number of city defences in the area under discussion. The intention is, however, to provide pointers towards the strategic value put upon urban defensive sites by the Roman military, and therefore to give some parallels which may be useful for those involved in the study of similar defences in Britain.

Apart from some exceptions near the frontier zones, most Roman walled cities in mainland Europe were fortified for the first time in the late 3rd and early 4th centuries, a period when barbarian raiding on a large scale had become a severe problem to be reckoned with by Roman military authorities. A very general, and simple, view is that virtually no cities or towns of purely civilian nature in Europe—this must exclude places which were actually legionary fortresses, like Mainz, Strasbourg, Regensburg, or León—were walled from the period after the death of Augustus until the mid or late 3rd century: thus there are early walls at Augustan colonies such as Arles, Nîmes, Vienne, Fréjus, Mérida, Barcelona, and Saragossa, but thereafter a great pause. Of towns and cities walled between the Augustan and the late Roman periods, the most interesting places for study are perhaps Heddernheim, Ladenburg, and Wimpfen, all of which lay beyond the Rhine in the Agri Decumates, land given up to barbarian pressure in the 260s. Elsewhere on the frontiers there was similar pressure. The number of civilian towns walled before this date is remarkably small—among the certain ones are Xanten, Cologne, Trier, Tongres, Wels, Augsburg, Avenches, and Aquincum civilian town (Budapest). Perhaps the most unusual are two towns whose walls, for some unexplained reason, were never finished. One is Augst whose walls cut off two of the approach routes to the city (from the south-east and west) and while they each contain a gate, it would have been easily possible to circumvent the walls (Laur-Belart 1966, 30f). Another such town is Lienz in Noricum Mediterraneum (*Jahreshefte des Österreichischen Archäologischen Instituts*, 1935, Beiblatt 1f: 1953, Beiblatt 93f). There are of course a number of cities whose walls are not known, or not securely dated—one might mention such important towns as Szombathely or Sremska Mitroviča—which may well have had defences in an earlier Roman period. As yet, however, there is nothing in the west to parallel the early boost to eastern town wall construction given by the Severan emperors, as shown on their coins struck to commemorate the construction of the walls at Nicopolis, Augusta Traiana, and other Thracian and Moesian cities (Hoddinott 1975, *passim*). A further impetus was given in this area by the raiding of the 250s, and one of the best preserved examples of city walls, at Iznik in a nearby area of Asia Minor also under pressure, is dated by inscriptions to the reign of Claudius II (*CIG* 3747–8; Schneider & Karnapp 1938).

It is as a response to raiding at a similar date in the west, too, that the majority of town and city wall building was undertaken. The invasions which most affected Gaul occurred in or around 260, at a time when Postumus had assumed the imperial office there. A further series of raids seems to have happened in 275–6, causing a desperate situation which was only righted by the energetic emperor Probus. The depth of penetration of those raids is shown by Eutropius, who relates (IX, 9) the sort of story one might expect from the Historia Augusta, that the 'tribes of the Franks savaged Gaul, and reached Spain. After devastating and nearly taking Tarraco apart, a portion of their force eventually grabbed some ships and even reached as far as Africa'. Other areas suffered raids at other times, but within the same span of 15–20 years after AD 260. The impact that the long and costly campaign of wall-building had on ancient historians and chroniclers is minimal. When one turns to tangible archaeological evidence, only two inscriptions recording the construction of walls or portions of them have survived from this period in the west. These are from Verona, recording construction in AD 265 (*CIL*, **5**, 3329), and Grenoble, recording the construction of the *muri Cularonenses* at the expense of Diocletian and Maximian in AD 286 or after (*CIL*, **12**, 2229). Apart from these, and a sentence in a letter by the emperor Julian (*Convivium*, 314b) many years later, admiring Probus for restoring 70 cities, one is clutching at straws to find anything concrete within the literary or epigraphic record. Such straws can be fragments of Gregory of Tours' *History of the Franks*, where, describing Dijon (**3**, 19), he mentions that the 'older folk say that the walls were built under Aurelian'—and even he spells Aurelian wrongly—or they can be entries in Amatus's *Dictionary of Geography*, which says that Pula, on the Istrian coast, had a Hercules and a Jupiter gate, thereby suggesting that these, like those of Grenoble, were built under the Tetrarchy.

The cost of building large numbers of city walls could not all be borne by the state, and there were various ways in which the problem posed for a local authority by the expenditure of so much capital could be alleviated. Those cities, like Grenoble, which seem to have had the construction paid for direct, were doubtless the lucky ones; to others the emperors might lend architects or detachments of military engineers to oversee the work; while others perhaps received some remission of taxes to enable progress to be made, whilst compelling all citizens to contribute, whether by providing cash or labour.

One may legitimately ask what the inhabitants gained from this arrangement. The majority of Gallic cities now began the construction of walls which encompassed normally at the greatest only a fifth, and sometimes as little as a tenth, of their original size. This can be shown most clearly from Paris (Fig 77), where the main centre of the former town was left as suburbs, and the only defended area was the Île de la Cité, which can have held few, if any, dwellings, since within its small space it included accommodation fit for the Caesar Julian when on campaign in the 350s. His troops had to bivouac, as Ammianus then says, 'in the suburbs' (XX,4,12; Julian, *Misopogon*, 340D–341D). All over Gaul, the same retrenchment was undertaken: at Périgueux, for example, the amphitheatre was used as a kind of vast forward bastion on which to hang the constructed circle of town walls (Fig 78). The same happened at Amiens and Tours and the amphitheatre was also incorporated within the walls at Rome, Arles, and Trier. Even cities with extensive walls, like Nîmes, Arles, Autun, Tongres, and possibly Avenches, had to cut down on the former extent of their defences. Perhaps the most surprising is Xanten, where, despite the Trajanic defences, what is known of the late Roman pattern suggests that on three, if not four, sides a completely different line was chosen (see Precht, this volume).

We are singularly ill-informed about what actually did lie within the walls of such diminutive cities. We have already seen what happened in Paris: the city or citadel area of most Gallic cities has been thoroughly built over

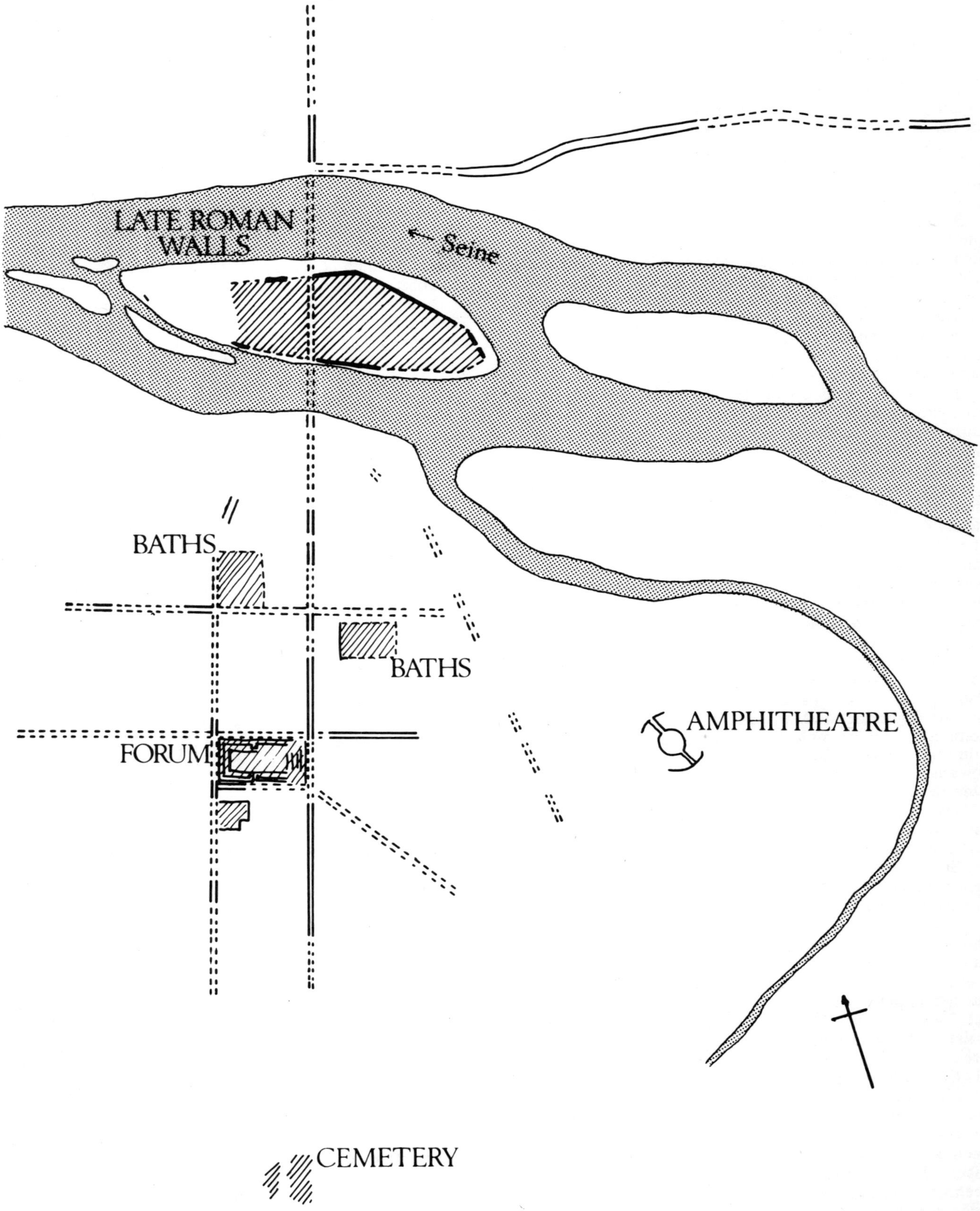

Fig 77 Paris: the extent of the 1st–3rd century city, and the late Roman defended enclosure of the Ile-de-Paris (scale approx 1:1470)

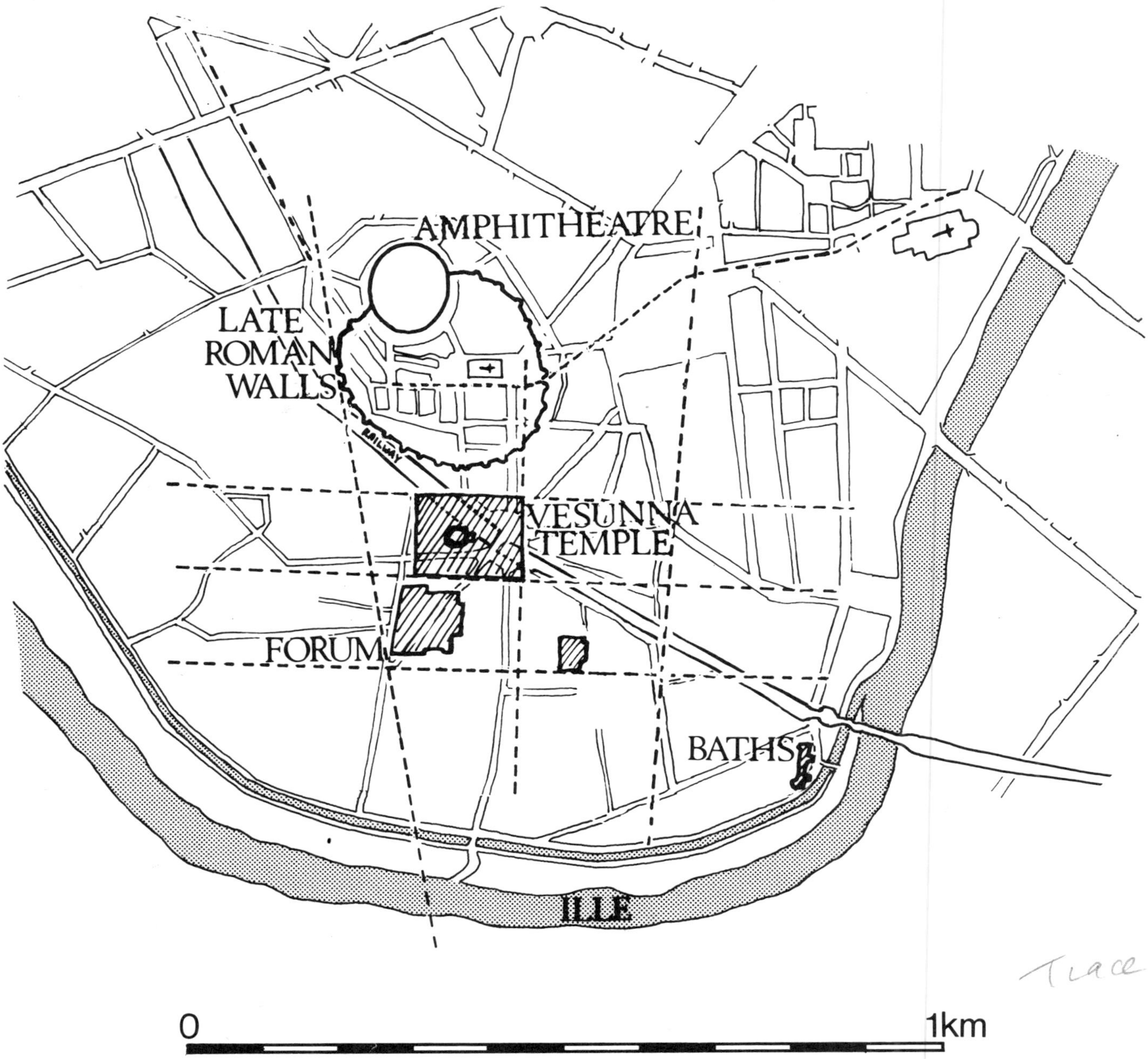

Fig 78 Périgueux: the extent of the 1st–3rd century city, and the late Roman defended enclosure attached to the amphitheatre (scale approx 1:1250)

in medieval and modern times. The chance of excavation within these inner city areas comes but rarely, and of course the possibilities of distinguishing 4th century structures from earlier ones are rarely realized. The best known example, at Bavai where the walls enclose little more than the forum or the *cryptoporticus,* is not a normal case; for Bavai, once the *caput* of the Nervii, had by the 3rd century lost that position. Exactly what the Bavai walls protected at this period is not known, but the walls are substantial and built in two separate building campaigns, though not precisely dated. Not all cities were quite so small: the cities of Sens and Bourges, for example, have a broad sweep of city walls which enclose 60 or so ha: these scarcely grew beyond the limits of their walls until after the end of the medieval period. Within such an area, there was adequate space for all the panoply of Roman life, including dwellings and tenement blocks. It is easy to see the development of the medieval town, with people crowding within the protective walls, stemming from the setting of the city limits by the construction of such solid defences. The small vignettes on the picture pages of the *Notitia Dignitatum*, or the city views on coins and medallions (including London, on the Arras Medallion used as the symbol for this conference—see cover) give some impression of the clusters of piled-up buildings within the city defences.

One way of understanding the reasons for the choice of a site for defensive treatment in the late 3rd and 4th centuries is to look at those towns and cities which did not survive the invasions by *barbari.* Assessment of this, too, relies heavily on the amount of excavation work at the site, but in most cases a low-lying town site was abandoned in order that a more easily defensible site in the neighbourhood (occasionally even the original hillfort precursor of the town) could be brought into use. This seems to have

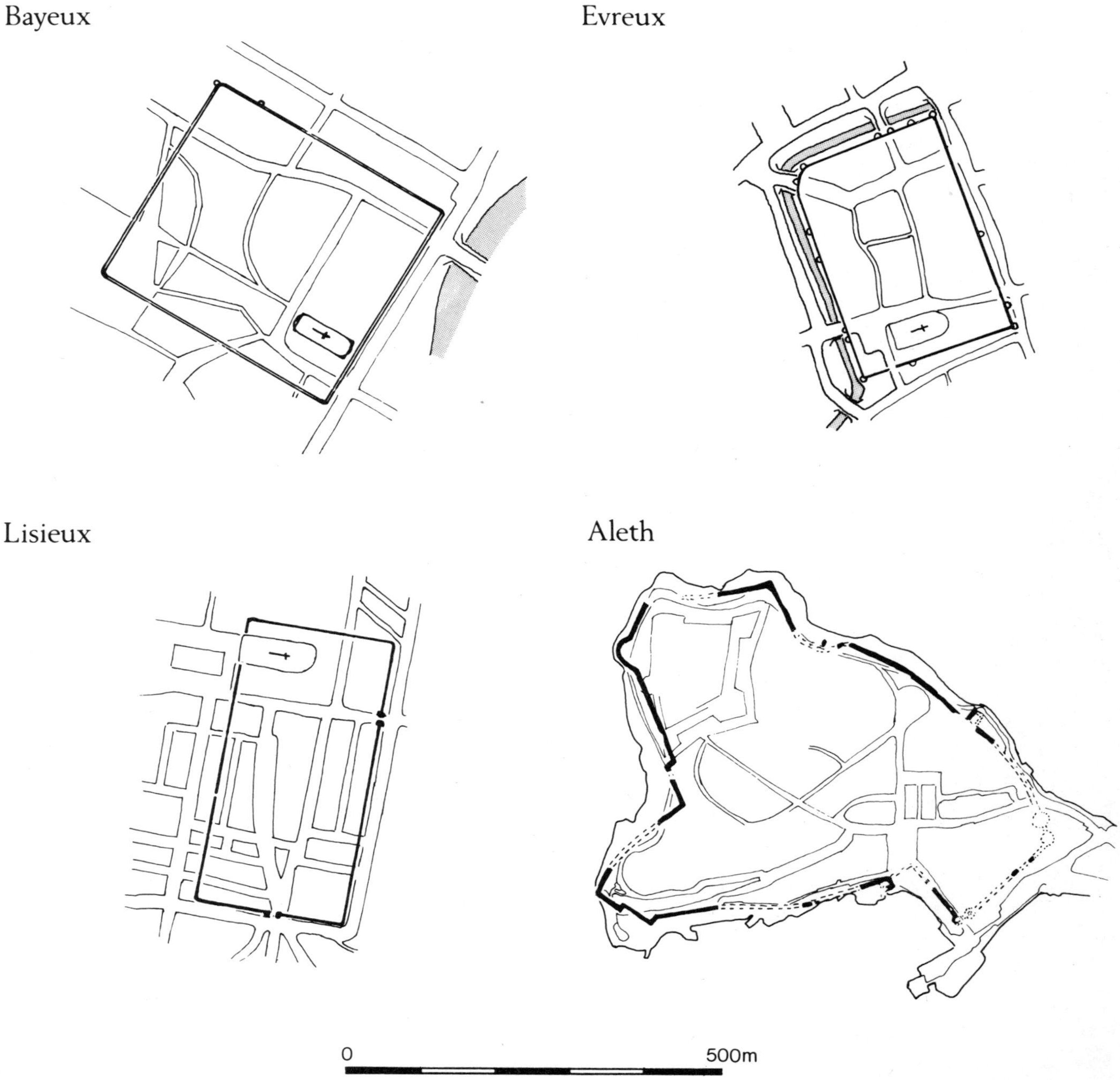

Fig 79 Late Roman walled sites in Lugdunensis II (scale 1:10,000)

been the case, for example, at Kempten, St Pölten (Virunum), and St Quentin, and probably at one of the most famous walled cities of them all, Carcassonne, though there will be other less well known instances.

This gives a stray hint that it was not merely local sentiment which caused the survival or even rebirth of a town at this date. We know (only from the *Scriptores Historiae Augustae*) that in Thrace, Gallienus sent teams of architects and engineers round the cities to organize their defences in the 260s (*SHA, Triginta Tyranni*, 23, 13, 6). The clear eye for strategic location and positional need evinced by the late Roman walled centres suggests more than local initiative. This is born out too by the groupings of walls into constructional types. I published a paper on this some years ago, attempting to define a group of city walls in Gallia Belgica which bore the architectural signature of a single hand or a single designer (Johnson 1973). The identification of such a group depends, of course, on distinguishing it from other groups in other areas, a factor which was not brought out in my earlier paper. A similar grouping, with evidence even more sketchy than the first, can be made in other areas too. In Lugdunensis II, for example (Fig 79), where the walled towns of Rouen, Lisieux, Bayeux, and Evreux lie, all the walls there constructed, apart from the coastal site of Aleth, are of rectangular ground plan with rounded corners, all have projecting semicircular towers, and at Lisieux in particular, this form of construction is placed at the foot of a substantial hill, where a more sensitive strategic design would have made more of the site's potential. In addition, the

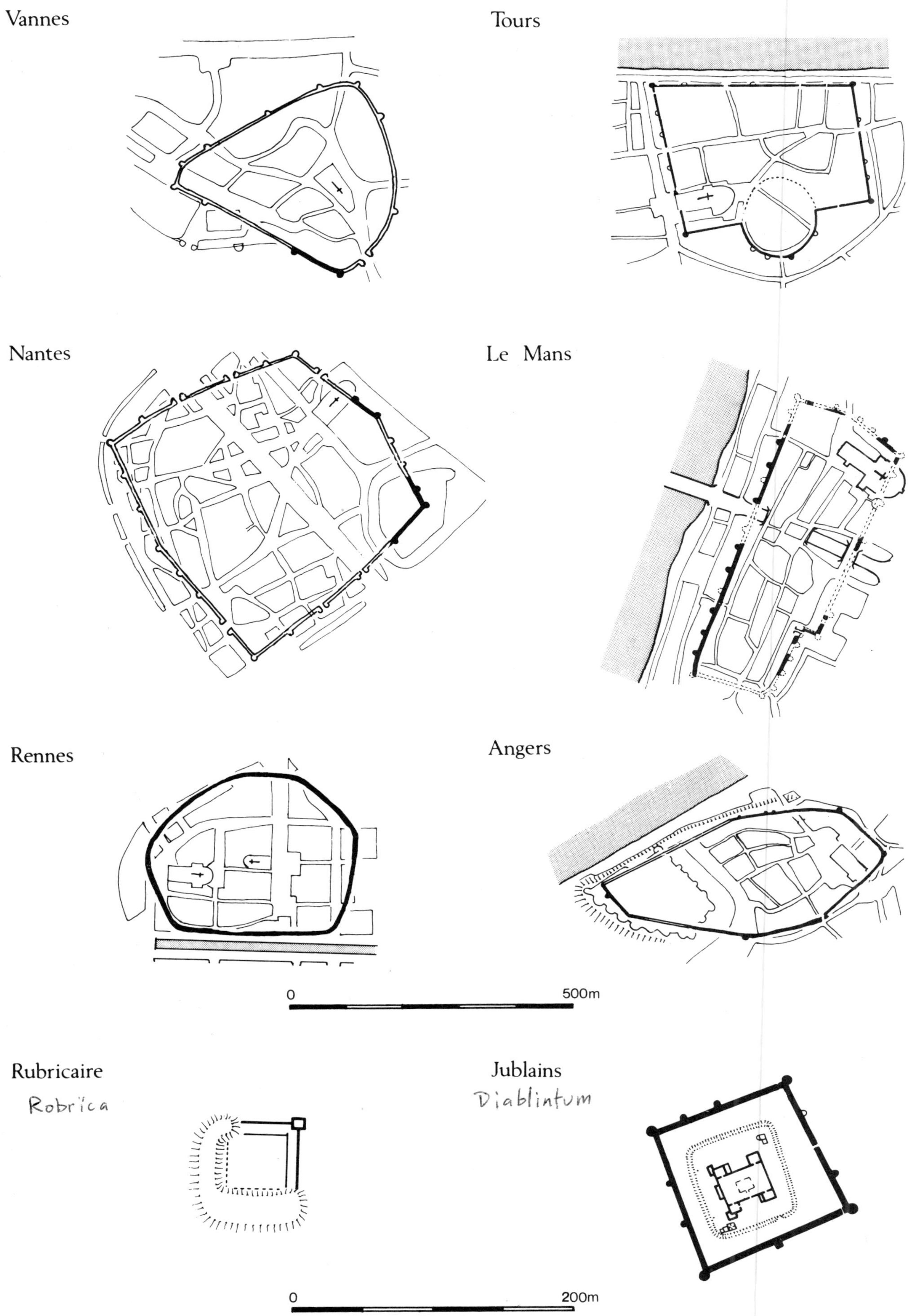

Fig 80 Late Roman walled sites in Lugdunensis III (scale 1:10,000, except Jublains and Rubricaire, scale 1:4000)

spacing and type of the wall-facing at all sites is similar: double tile-courses are found with a large expanse of small blockwork between them.

Further west in Lugdunensis III (Fig 80), the *civitas* walls have certain features in common. Here the enclosed areas are more irregular, but their style of building is very similar. In complete contrast with what we find in Belgica II the facing uses many tiles in thick, closely-spaced courses, as at Le Mans and Rennes, which became known as the *Ville Rouge* in the Middle Ages. In addition, the putlogs are usually found set centrally in the tile-courses, and not in the course of *petit appareil* above them as at Beauvais. At Le Mans and Rennes too, there was also patterning in the walls, made by using ironstone and normal limestone in geometric designs. At Le Mans various building gangs displayed their own distinctive style of decoration in various parts of the walls (Butler 1958). The sizes of materials used and the similarity of the tower types in the area point to another gang of military builders employed to build the defences of this group of *civitates*, which includes Le Mans, Rennes, Angers, and Nantes.

A fourth distinctive group of *civitates* with similar defences lies in Novempopulana (Fig 81). Among these it is not so much a correspondence of sizes and details of construction which provide the closest points of comparison, but the overall size of the areas enclosed by the walls and the fact that most of them occupy very similar types of site on the tops of small but prominent hills, as instanced by St Bertrand de Comminges, Lescar, St Lizier, and Oloron. There is some rough correspondence between the sizes of the small blockwork used, and in the fact that the general building styles of these small defended areas are roughly similar. All in all, it appears that a single gang was also responsible for this series of defences, just as other sites in other areas of Gaul show a certain interrelation, suggesting a unified plan in their construction.

If, then, these *civitas* defences do form cohesive groups built by the same builders, then any evidence (which in general is in very short supply among Gallic *civitates*) for dating one of them can be legitimately extended, within careful limits, to the remainder of the cities in their group. The builders were presumably working on a contemporaneous programme, and will have gone round from one city to another within the group. The pooling of evidence from a number of *civitates* in a province can produce some interesting results: the dating of Nantes and Rennes, for example, is based on a cache of milestones of Victorinus and Tetricus. There are so many at Nantes and Rennes that it is perhaps safe to assume a 'hoard' dating of *c* 275. All other cities in the same area, if built by the same troops, can be dated also to the same period. A similar dating can be attempted for the *civitates* of Belgica Secunda, where coins of Postumus and Diocletian have been found in the wall-core of Beauvais: this can be extended to the other cities in the region to assume that they were all constructed in the period round 285.

The identification of groupings of such walls can be carried outside Gaul. As was recognized 50 years ago by Sir Ian Richmond, there is a group of late Roman towns in Spain, all apparently of very similar build (Richmond 1931). They include the impressive large walled centres of Lugo, Astorga, and León, the last of which was actually a legionary fortress of Trajanic date whose walls were strengthened in the later Roman period. An extra skin of masonry was added on the exterior, constructed of large blocks of *opus quadratum*, and incorporating the kind of large projecting towers which are familiar from other Roman towns in western Europe. These are also visible at Astorga and at Lugo, perhaps one of the best surviving stretches of late Roman walls anywhere in Europe. Other towns in Spain—Barcelona, Saragossa, and Iruña—and some also in Portugal—Evora, Béjo, Coria, and Condeixha-a-Velha—have walls in similar style (Balil 1960), but the group of defences in the north-west corner of Spain is a remarkable concentration, based on the fortress of León and probably a deliberate move to protect the lucrative gold mines, particularly important to the Roman economy after the final abandonment of Dacia by Aurelian in the 270s.

At the other end of the western empire, in Pannonia, there lies a group of walled centres which form the nodal points of the road network. Though in some cases imperfectly known, most of these centres are of late Roman date. Most seem to be rectangular areas, and the best known, Fenékpuszta, has a very curious collection of separate buildings within it (Mocsy 1974, 302f). This is, however, only one of several such posts, including Környe, Ságvár, Kisárpás, and Alsóheténypuszta, which has been archaeologically examined in the relatively recent past (Soproni 1978, 138–55). Here it was discovered that a large rectangular area was defended by strong walls with alternate U-shaped and round towers; there was no examination of interior buildings. The sites of these 'towns', if I may call them thus, are not in any outstanding defensible strategic locations, and several have late Roman cemeteries containing military graves outside their walls. Dr Sándor Soproni, the excavator of the site at Alsóheténypuszta, has suggested that these sites are in fact of military origin and held a military garrison and provided support bases for the frontier troops along the Danube bank. The large walled areas, only half filled with buildings, will also have served as refuges for the local population in times of danger. Although these centres may not be towns, they provide an example of a type of fortification to watch out for. Their dating is assigned to the Constantinian period on the basis of the U-shaped towers, replaced under Constantius II by round ones.

We return, finally, to the towns themselves. We have seen how serious the barbarian threat of the mid to late 3rd century had seemed to the unwalled European Roman towns often deep inside the largely unprotected provinces. The new defences constructed at this period enclosed on the whole a diminutive area, and study in detail of their building techniques has suggested that there were groups of military architects or builders working in distinctive styles within restricted areas. The identification of such groups of defences leads to the supposition that strategic and tactical reasons called for the walling of small heavily barricaded posts at this time—administrative centres only, in the main, not areas for expanded gracious living. Such walled centres seemed to the Roman military commanders the minimum necessary requirement for the preservation of the Roman administrative machine in the face of mounting hostile pressure from outside; in the case of the ring of Spanish sites protecting the gold mines, banditry from within as well as hostile pressure from outside the empire doubtless justified what seem to be the excessive defensive measures.

This is not to claim that every late Roman defended city, town, or road post was part of a military grand strategy. Technically speaking, the emperor's permission was required for any city wishing to build defences, and it is hard to see that at this period permission would have been refused; doubtless also the civil service was occasionally asked to ratify the construction of defences after the event. What is not known, and is perhaps more interesting, is how a particular city managed to achieve imperial financial support for wall construction. This is a subject too involved

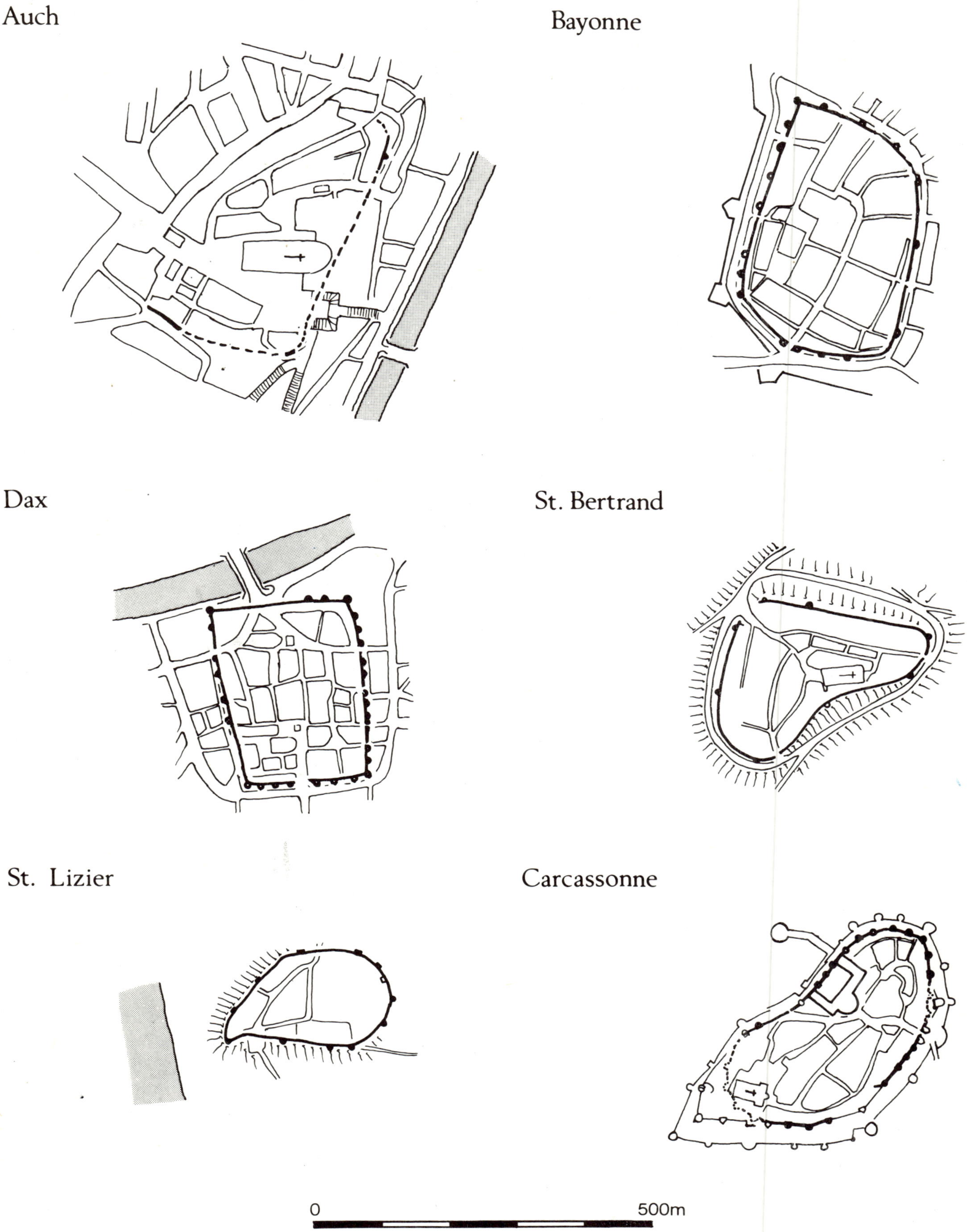

Fig 81 Late Roman walled sites in Novempopulana (scale 1:10,000)

to go into here, for it entails an understanding of power struggles within the imperial court. Suffice it to say that in the study of Roman urban defences, the construction of each particular city's set of walls is not necessarily a local phenomenon, and that even within such a large and diverse subject, where dating is particularly hard to come by, one can occasionally see signs of an imperial or military pattern emerging.

References

Balil, A, 1960 La defensa de Hispania en el Baio Impero, *Zephyrus*, **11**, 179–97

Butler, R M, 1958 The Roman town walls of Le Mans, *J Roman Stud*, **48**, 33–9

CIG *Corpus Inscriptionum Graecarum*

CIL *Corpus Inscriptionum Latinarum*

Hoddinott, R F, 1975 *Bulgaria in antiquity*

Johnson, J S, 1973 A group of late Roman city walls in Gallia Belgica, *Britannia*, **4**, 210–23

Laur-Belart, R, 1966 *Führer durch Augusta Raurica*, 4 edn, Basel

Mocsy, A, 1974 *Pannonia and Upper Moesia*

Richmond, I A, 1931 Five town walls in Hispania Citerior, *J Roman Stud*, **21**, 86–100

Schneider, A M, & Karnapp, W, 1938 *Die Stadtmauer von Iznik*

Soproni, S, 1978 *Der spätrömischen limes zwischen Esztergom und Szentendre*, Budapest

SHA *Scriptores Historiae Augustae*

Fig V Britain

PART II BRITAIN

Roman urban defences: a review of research in Britain

Brian Hobley

The methodical investigation of Romano-British town defences began some 50 years ago and has steadily grown in importance, being generally accepted by scholars as vital to an understanding of the history of the province (Frere 1967, 250).[1] Urban settlements ranging from formal chartered cities to large villages were introduced to Britain shortly after the conquest in AD 43, and their archaeological investigation is required to gain the fullest possible knowledge of their birth, growth, and decline. Much of this understanding can be provided by the study of defences, which in many cases formed the boundaries of the settlements. Research has shown how varied cities and settlements could be in their origins, development, economy, markets, house plans, and defences. No two towns in Roman Britain were alike. Significantly, Rivet has stressed (1975, 113) that 'we were wrong . . . to call everything that had a wall a "town": other considerations apart, it was one of the glories of the early Roman Empire that it created conditions in which towns without walls were a possibility'. As late as 1969 the idea that all Romano-British towns possessed defences continued to be held (Collingwood & Richmond 1969, 100). Consequently, attention is now being focused not only on the reasons for defences, ie civic pride, defence expediency, special local circumstances, and definition of town limits, but also on the criteria which allowed some civil settlements to build defences and others not (Rodwell & Rowley 1975). On this aspect of research both Wacher (1966, 67) and Webster (1975, 56) have highlighted the unfortunate tendency to regard defences as a separate entity, and the consequent dangers of losing sight of the town as a whole: a careful study of the internal and external buildings may provide valuable clues to the understanding of town fortifications.

An important aspect central to any consideration of why defences were built is whether or not authority was vested solely in the provincial or imperial government. However, to date, the economic strength of the local community, especially that of wealthy patrons, and its ability to pay for defences—particularly in the 3rd century—has not been fully considered. By this date a sound local economy based on a thriving hinterland was vital for the survival of any town in the north-west provinces. Many towns, however, may have been too poor to pay for the construction of walls, especially if they were close together and competing for the wealth of the hinterland.

Today's received wisdom would allow that, at least in the early empire, town defences could seldom have been built for local reasons. Frere has stated, 'Since urban defences are not random phenomena but were subject to central decision they form a class in which the date of any one will assist the dating of the others' (Frere 1965, 138). In other words, they are more likely to be contemporary than of differing dates. If this premise becomes untenable, many conclusions hitherto reached about town defences will have to be radically revised. The simple fact that cannot be emphasized enough is that, notwithstanding more than half a century of investigation, not a single town of Roman Britain has had its defences fully investigated or satisfactorily explained, and therefore Frere's hypothesis cannot yet be rigorously tested.

Philip Corder, inspired by Ian Richmond and his work on the walls of Rome, undertook research in the 1950s which produced seminal results and has shaped methods of approach to this enquiry ever since. He observed that 'the interpretation of the evidence has been influenced by prevailing fashion or has relied too much on the preconceived notions of the ancient historian' (Corder 1955, 21); and again, as if to stress his fear for future studies, he made a plea for 'independence of outlook in its interpretation' (*op cit*, 22). Unhappily, this excellent advice has not been heeded and the tentative nature of the conclusions of his own paper has tended to become enshrined as historical fact.

John Wacher in his equally influential paper on 2nd century earthwork defences (Wacher 1962) also stated that his conclusions were tentative and that much more evidence was required before they could be accepted as more than hypothesis; again much stress was put on the limitations of the evidence. Michael Jarrett, with characteristic directness, asserted that attempts to produce a definitive statement of any kind on Romano-British defences were premature (Jarrett 1965). The stimulating exchange of views in *Antiquity* during 1965 between Jarrett, Frere, and Wacher is required reading for any who doubt the complexities of this field of investigation, especially concerning the dating evidence (*Antiquity*, **39,** 57, 137, 225). Jarrett saw how easy it was for a working hypothesis to become accepted dogma, and he sought to prevent the creation of simplistic frameworks until 'the quantity and quality of the evidence would permit it'. Moreover, he argued that 'each town must be considered on the evidence it produces . . . . The attempt to illumine one town by reference to another serves only to darken the whole' (Jarrett 1965, 227). The diversity of approach to urban defence studies was thus polarized by Frere and Jarrett, with few heeding the words of Corder or Wacher regarding the tentative nature of their conclusions. Consequently, the pendulum of fashion governing prevailing hypotheses on town defences has been swinging from the 1930s to the present day, apparently kept in motion by the tension spring of a need for a chronological framework, however inaccurate this makes individual clocks.

The results of one of the earliest scientific excavations of a Roman town in Britain—at Verulamium, between 1930 and 1934, which led Wheeler to suggest a Hadrianic date in the 120s for the construction of the town wall and gates—became the first major influence on the dating of urban defences (Wheeler & Wheeler 1936). However, Wheeler mistakenly equated the date of the construction with that of the associated artefacts, when in fact they pointed to a *terminus post quem*—a date some time after that indicated by the samian pottery and coin evidence (*ibid*, 58). Furthermore, the coarse pottery was not taken into consideration, and the samian is now viewed as being mid 2nd century. This report was certainly the first to interpret the results gained in the wider context of the history of the Roman Empire. However, such was Wheeler's influence that it became fashionable to ascribe a similar date to Roman towns elsewhere in Britain. R G Collingwood accepted Wheeler's results from Verulamium and used them as a framework for the history of urbanization in Britain, and as a whole they have coloured popular history ever since.

In the early 1950s Corder excavated both fort and civil settlement at Great Casterton and half a dozen sections were cut through the defences primarily for dating evidence, with equally influential results (Corder 1955). After surveying results from other towns he concluded that many town defences had been built towards the end of the 2nd century. Regardless of the author's request for caution, another fashionable framework was thereby adopted. Wacher followed and strove to place the rash of 2nd century earthwork defences into an historical context, recognizing a south-west group with ramparts and an eastern group without (Wacher 1962). Though history is silent about the years in question, he believed that trouble in Wales and a depleted army in that area allowed serious unrest to develop which finally led to a tribal rebellion. With certain modifications this theory has survived well and is still part of current thinking, though it is not without its critics, especially as to the accuracy of the dating of these earthworks (Rivet 1966, 105–6). Such is the background to the inherent dangers of unquestioning acceptance of theories about town defences which have become fashionable, illustrating the need to learn now from the mistakes of the past and to interpret more cautiously.

Notwithstanding this cautionary approach to the existing evidence, four main recognizable types of linear town defences can now be discussed.

Initially, in the 1st century, a relatively small number of civil sites was provided with earth and turf ramparts. Then, in the second half of the 2nd century, many more settlements received earthwork fortifications. Early in the 3rd century and possibly by the late 2nd, town walls were built into the front of the earthworks. Finally, in the second half of the 4th century, bastions or projecting towers were added to a number of town walls, accompanied by wide, flat-bottomed ditches dug to replace the earlier narrower ditches.

This outline of developments in Britain should now be related to historical events on the continent. The great crisis of the second half of the 3rd century, when between 230 and 260 Rome lost her military initiative to the barbarian world of north-west Europe, resulted in radical changes in defensive architecture. Before this date both fort and town defences were conceived as playing a passive role and a limited use of the various materials available produced a simple architectural design. The whole appearance could be said to reflect the army's ability to contain any real threat to the towns at the frontier or outside the empire (see Johnson, this volume). Before the time of Constantine, no forces were permanently deployed in large concentrations behind the frontier. Town defences, therefore, were essentially designed as protection against occasional local troubles such as banditry and small-scale uprisings. Structural details and materials were simple and modelled on those used by military architects, and in some cases turf and timber were used, which were only slowly replaced by stone in the 2nd century. Curtain walls of stone tended to be of minimum thickness and gates were simple in design, though with some architectural elaboration to gratify local pride. The barbarian invasions saw the defences of both forts and towns replaced in a manner which was in direct contrast with what had gone before; they were rebuilt on a truly massive scale in the thickness and the height of walls and projecting towers and the size of the ditches. The quite different, if not unique, nature of the development of urban fortifications in Britain must be seen against this continental background.

In 1930 the investigation of Romano-British town defences, with the exception of Colchester and London, had not yet commenced (Wheeler 1928, 69–106). By 1955, when Corder formulated his hypothesis, only 22 sites had been investigated in some form; these included Verulamium, Colchester, Lincoln, Gloucester, Silchester, Caerwent, Exeter, Leicester, Winchester, Canterbury, Cirencester, Caistor-by-Norwich, Dorchester (Oxfordshire), Dorchester (Dorset), Alchester, Chichester, Towcester, Mancetter, Aldborough, Brough, and Great Casterton. By the end of 1979 few of the 150 settlements listed in the 1978 Ordnance Survey map of Roman Britain (Ordnance Survey 1978) that are relevant to this study had not been to some extent investigated by aerial photography, field walking, or excavation. It must be stressed immediately that not all these settlements were found to have defences (Webster 1975). Nevertheless, it was vital that they should be investigated before conclusions could be reached about the nature both of town defences and of urban settlements without defences.

Current ideas on Romano-British town defences suggest that the first major provision came towards the end of the 2nd century, when many towns received earth ramparts, a provision, it should be noted, which is apparently without parallel in any other province of the Empire (Frere 1978, 285). Before that date, it appears that the towns of Britain, in the main, did not have fortifications, in keeping with the broad imperial strategy of frontier defence. However, the three *coloniae* of Colchester (Frere 1978, 283), Gloucester, and Lincoln (see Jones, this volume) were certainly provided with narrow-gauge stone walls, possibly around the late 1st century. Otherwise, earthwork defences have been recognized at the *civitas* capitals of Silchester (see Fulford, this volume), Verulamium (*Antiq J*, 1960, 40; 1961, 41, 80, 82; Frere 1964a, 65; 1964b, 104), and probably Winchester (*Antiq J*, 1970, 284). At Cambridge (*Britannia*, 1976, 340), Godmanchester (*Britannia*, 1975, 250), Neatham (*Britannia*, 1975, 6, 215, 278), and Ilchester (*Britannia*, 1975, 275) there is also some structural evidence for 1st century defences, though the dating is somewhat uncertain. Whereas there is as yet no evidence that any of the villages were fortified at this time, the enigmatic defended road stations to the east of Chichester on Stane Street at Hardham (Cunliffe 1973, 69–72) and Alfoldean (Wacher 1978, 96; Cunliffe 1973, 71–2), are believed to be of 1st century date, though accurate dating is lacking (Fig 82a).

It has been suggested that, since the provision of defences at this time for a small number of sites is somewhat exceptional and anomalous, special explanations have to be sought. The *coloniae* defences could be explained by civic pride in being chartered cities, likewise Verulamium in its status as a *municipium* and Silchester as being within the client kingdom of Cogidubnus. A major threat at this time would presumably have justified a far greater number of towns being defended than is at present known. For example, the early Claudian defences at Silchester, Verulamium, and Winchester could be seen as a consequence of the guerrilla warfare of Caratacus, since they were very close to the frontier based on the Fosse Way. The Boudican rebellion and its aftermath could certainly be considered a valid reason for defences for these three towns, to which can now be added the early defensive ditch at Colchester. In addition to these early events justifying defences, the effects of the civil war of 68–69 within the kingdom of Cogidubnus, which may have been responsible for the establishment of defended road stations to protect the imperial post, have also to be considered.

The omission of other towns, principally London, from

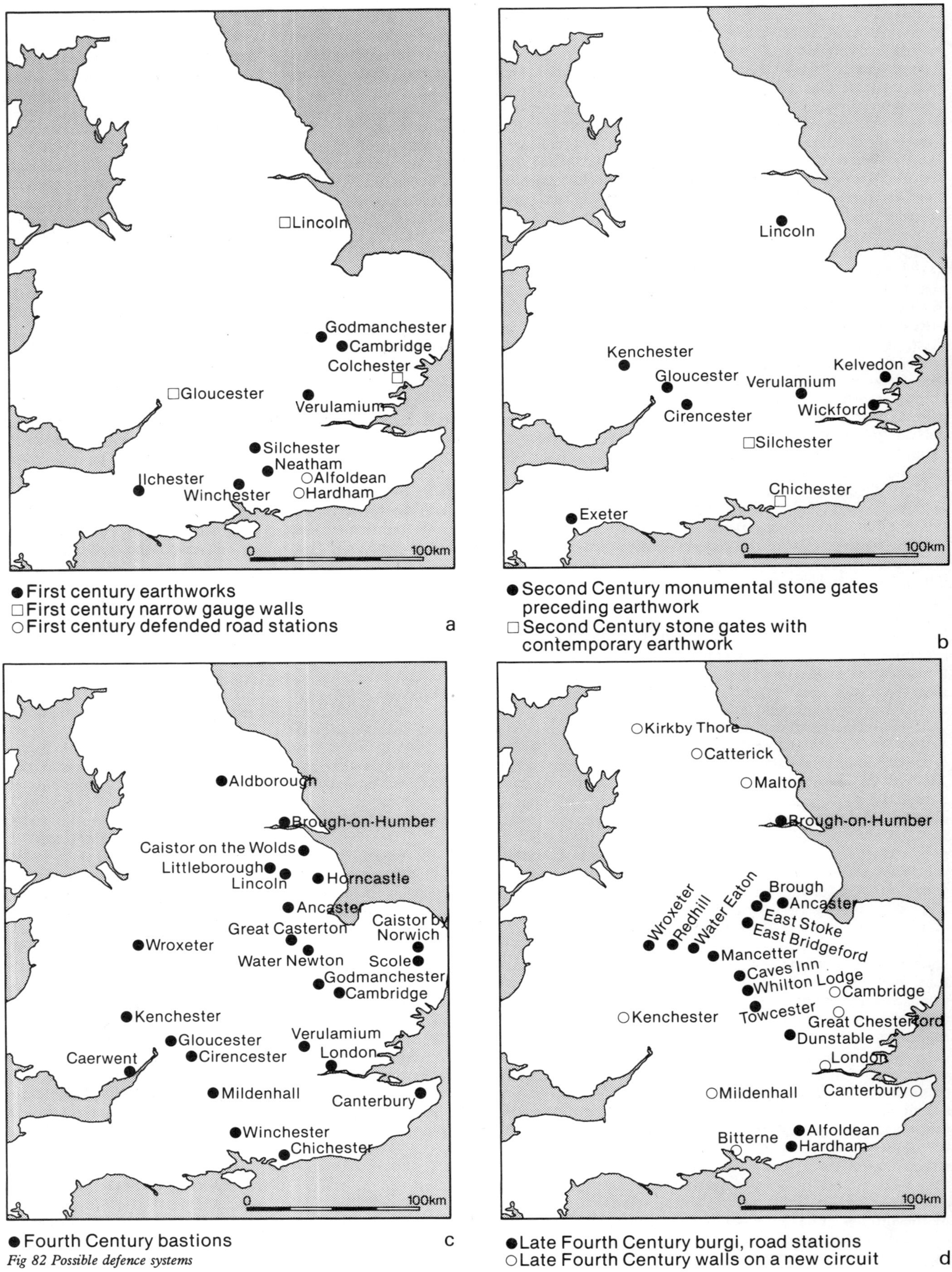

Fig 82 Possible defence systems

these protective measures is difficult to explain, but it may be, as at Verulamium until the excavations of the 1950s, that other early defensive circuits have yet to be discovered. If such proves to be the case, it will have to be considered whether there was a more extensive provision of town defences. In addition, an explanation is still required for the relevant but somewhat inconclusive evidence for late 1st century to mid 2nd century earthwork defences in the south-east at a number of towns, including Exeter, Chelmsford, Dorchester-on-Thames, Godmanchester, Ilchester, Kelvedon, and Worcester, and others in the Midlands and north at Brough-on-Humber and Mancetter (Wacher 1962).

By the end of the 2nd century, earthwork town defences were widespread, notable exceptions being Canterbury, Colchester, Leicester, and London. At Verulamium the Fosse earthwork of mid–late 2nd century date has been shown by Frere to have been unfinished (Frere 1972, 108). Wheeler's observation of additions to the bank is now unacceptable as evidence for two phases, since, as Frere has suggested, it is unlikely that additions would have been made to an unfinished circuit.

Significantly, a number of gateways were built, mainly at *civitas* capitals, which apparently preceded the earthworks, eg at Exeter, Cirencester, Verulamium, and the *coloniae* Gloucester and Lincoln. However, the gates at Silchester are now known to be contemporary with the earthworks (Fig 82b; and see Fulford, this volume).

The historical background to this activity is thought to be the period 193–196/7 when the governor of Britain, Clodius Albinus, made his unsuccessful bid for the imperial purple. However, since the earthwork defences are not closely dated, another, unknown, major historical event may be the cause. Frere and Wacher have both argued that this activity was the 'application of a single policy in a single context which demanded the rapid simultaneous fortification of all sizeable settlements' (Frere 1978, 285). If the Albinus episode was the relevant historical context, the amount of work undertaken in such a short period of time was quite prodigious. Moreover, the stone gateways have of recent years been interpreted as evidence for an earlier commitment to the construction of defences in stone; when this plan had to be changed, building materials that could be used quickly by unskilled labour had to be substituted. At Verulamium straight joints between the wall and gates clearly show that most of the masonry gates antedate the wall and are very probably contemporary with the earthbank, an important detail overlooked by Wheeler. Moreover, there is an additional point of interest in that the easternmost monumental free-standing arch discovered by Wheeler was constructed when the city boundary ditch was dug. The western arch is mid 3rd century in date and therefore contemporary with the wall.

If some stone gateways were pre-Albinian in date, it would go some way to explaining the monumental gates at Colchester, Lincoln, and elsewhere, the erection of which is difficult to reconcile with the emergency situation facing Albinus. In this connection, the singularly impressive gate at Cirencester is unusual if not unique in the Roman world in having four carriageways. At Winchester and Lincoln, the unusual practice of rebuilding in stone over earlier late 1st century timber gates has been recorded. At London, the Roman gate at Newgate appears to be later than the wall for the gate is 4½ ft higher in ground level than the wall. Conversely, other gates in London, eg at Ludgate and Aldgate, appear to be earlier than the wall, as the foundations are of the same flint and puddled clay construction as that used in the Flavian forum.

The provision of a stone wall some 2·4 m wide and nearly two miles long appears to indicate that London was singled out for special treatment at this date, presumably because Albinus needed the strongest possible defence for his capital against the contingency of a rapid retreat from the continent. But its size and length, and the use of stone from distant quarries, may have worked against completion within the six years Albinus was in control.

It is difficult to believe on economic grounds alone that Septimius Severus had the resources to circumvallate the towns of Britain as well as to restore the northern frontier. Further, on political grounds, it has to be appreciated that he was restoring to the Empire a rebellious province which should clearly not have stood to gain so immediately from large-scale public works. Caracalla (AD 211–17), who succeeded Severus and granted Roman citizenship in 212 to all free inhabitants of the Empire, could have continued the programme of walling (Rivet 1964, 93; Frere 1978, 288).

There is, however, general support for the view that this intensive fortification in Britain came earlier than for the towns of Gaul (Butler 1959; Rivet 1975, 105). Notwithstanding the problem of accurate dating both on the Continent and in Britain, it is widely agreed that town walls in Britain belong to an earlier style of 2nd century fort defences, while in Gaul town walls were built in an apparently later style and are dated just before the reign of Diocletian (284–305). By the early 3rd century in Britain a large number of towns possessed extensive stone defences and only a few, eg Chelmsford and Brampton (Norfolk), were excluded. Was their construction a reaction to events on the frontiers of the province? In the absence of conclusive evidence, both historical and archaeological, for walled circuits, attempts have been made to establish a chronology for stone defences based on architectural features, gates, internal towers, and bastions (Frere 1978, 286–7, 289, 291; Wacher 1975a, 75–7, 114; 1978, 99–100).

The establishment of three basic classes of town walls can be attempted:

1 Those with a bank which predates the wall and which has been cut away to insert the wall, so that the same ditch could continue in use
2 Those with a bank which has been piled up against the wall as it was built, therefore being contemporary
3 Those with no bank

This impressive building of masonry walls around both towns and villages probably lasted from *c* 200 to the 4th century, beginning with the walls of London *c* 190–210, Silchester *c* 225, Mildenhall *c* 360, and recently revised Kenchester and Caerwent *c* 350 (Wacher 1975b, 51; Frere 1978, 286–7, 289). However, the simple fact is that though stone walls generally postdate the Albinian episode, the major stone elements of defences, gates, and internal towers, and that of the narrow curtain walls, had already been introduced and belong to the older tradition. Corder argued from a *terminus post quem* pottery date that some walls which succeeded earth ramparts were built during the period of Clodius Albinus, but today, as mentioned, this is seen to be the main period for earthwork defences. At Verulamium a hoard of five coins concealed in the floor of one of the internal towers is unlikely to have been buried later than 240 and supports this date (Frere 1978, 287).

The Verulamium circuit, it should be noted, though planned on a different line, joined the gateways of the unfinished Antonine 'Fosse' earthwork and the defended bank of the river Ver for the first time. It also reduced the city from 93 to 81 ha.

In the mid 1930s at Verulamium Wheeler had seen the 3rd century as a period of economic crisis: 'Verulamium must at this time have borne some resemblance to a bombarded city' (Wheeler 1936, 28); and again, 'it is certainly true that virtually no building took place in the city, in the 3rd century' (Frere 1964b, 108). It is now known from excavations in the 1950s that Verulamium was in fact quite prosperous in the early 3rd century with large houses being built; in Insula XIV six substantial shops were built in the late 3rd century (Wacher 1975a, 218; Frere 1972, 98, 110). Thus the archaeological evidence now argues against a decline, though elsewhere in Britain it still holds good.

At Great Casterton, the first section led Corder to challenge the earlier Wheeler model of a Hadrianic date. From a personal account given to the writer, it seems that the principal excavators at Great Casterton had conflicting views regarding the validity of the Albinian episode as the only historical reason for the town's defences. Unfortunately, the excavators failed to detect that the wall was a later insertion into an earlier rampart. Today, these walls, like many others, are attributed to Septimius Severus at the beginning of the 3rd century, though Caracalla could also have planned the work, as it was his grant of citizenship to all freeborn men within the Empire that led to towns gaining in importance from the twin concepts of protection and civic pride.

Graham Webster has made a practical observation (pers comm) that by the 3rd century the 2nd century earth ramparts would have been rapidly eroding or collapsing into the ditch, with maintenance becoming increasingly difficult. Thus it need not have been the response to an immediate threat which motivated the construction of town walls in every case, but more the desire to retain an effective defensive circuit plus the realization that future maintenance would be easier and that the general appearance would gratify civic pride. The insertion of a wall into a decaying rampart would therefore have been the most sensible solution. Webster further suggests that priorities may have been established according to which ramparts were most eroded; hence the variation of dates for town walls.

Thus at present, the favoured date for most stone walls lies between 210 and 270, though many more exceptions are now becoming apparent; eg Canterbury and Caistor-by-Norwich were first defended late in the 3rd or early in the 4th century. In sum, the evidence would now begin to suggest a longer programme of work than earlier envisaged, which went on through the 3rd century as resources became available town by town, with private patronage undoubtedly affecting the pattern of building. Brian Hartley has stated (1966, 57), 'we may now be sure that some town walls were built very much later than used to be thought'. Some walls were certainly being built after 270, and at this time towns with a government function would also have been given high priority, especially if they formed part of a communication system.

One remarkable aspect of town walls in Roman Britain is that they virtually never contain reused masonry. Rare exceptions are the very late walls of the lower colony at Lincoln and the riverside wall at London which both therefore probably have to a degree the appearance of Romano-Gallic walls. In Gaul some 80 towns were provided with walls which on the evidence of much reused stone were said to be the results of panic and crisis. However, Stephen Johnson has noted detailed tooling and moulding on a number of those town walls, which hints at a favourable time factor, as does the observation that three towns appear to have had their walls constructed by the same builders, as yet an unrecorded practice in Britain (Johnson 1973, 217; see also Johnson, this volume). Further, in towns in Britain, unlike those in Gaul, the walled areas evidently either comprised the whole town or excluded only outlying parts.

If the majority of town walls in Roman Britain were built before 270, they antedate the town walls in Gaul and the series of Saxon Shore forts of the reign of Carausius (287–93). Johnson is inclined to the view that certain settlements were integrated with the Saxon Shore forts to produce groups of defended towns and villages with the forts acting as their strategic centres. Research for this paper has shown the possibility of such previously unrecognized chains of fortifications. Firstly, in the lower Severn valley, should Caerwent, Kenchester, and Mildenhall prove to have been walled *c* 350, they could prove to be part of such a system. Secondly, the provision of bastions in the second half of the 4th century could also be an indicator of a similar system. Unfortunately, as yet, the full extent of the remodelling of walls by the addition of projecting towers is not sufficiently known, but there is a strong indication that towns on the road south through Lincoln and Ancaster to Godmanchester were thus chosen. This may also have happened in the south-west, where a similar system of bastion defences would seem to have been needed (Fig 82c).

There is little direct evidence of how bastions in Britain functioned, but in Gaul, where they have survived better, they are equipped with 'catapult ports' or windows for the passage of projectiles, eg from *ballistae* or by archers. In Britain, Caerwent undoubtedly affords the best example of a bastioned town wall, but it is only at Pevensey that a bastion window has survived in part. The earliest British bastions at Verulamium are believed by Frere to have been a failure due to their being strategically badly placed on a gradual curve of the corner as well as projecting insufficiently to give flanking covering fire. However, Webster (pers comm) sees these towers as having been built primarily to provide forward fire power.

The addition of bastions to walled circuits required that a new ditch be dug further out between 60 and 100 ft wide, after the old ditch had been backfilled and the bastions constructed. A rare survival in London of this early ditch is thought to have been found at Ludgate. Usually, the later medieval ditch destroyed both the early and late Roman ditches.

It was Corder working at Great Casterton who first postulated the use of artillery weapons on the bastions in Britain, inspired by Richmond's work on the walls of Rome. The highly important discovery of the late 4th century date of the bastions clouded the recognition that the wall itself was an insertion into an earlier rampart. Unfortunately, the excavators believed (pers comm) that the discovery of the bastions was 'a bold enough step forward in itself'. Sadly for Roman town studies, had discoveries of the inserted wall and the later added bastions both been made at the time of excavation, a remarkable stride forward would have been achieved.

Bastions, however, were by no means commonly provided even for walls being newly built at this time, eg at Catterick and Great Chesterford, and probably at Cambridge. When built, bastions do not appear to have been provided for the whole of any walled circuit, unlike in Gaul. For example, at Caerwent bastions are only to be found on the south and north sides, which suggests an incomplete system. However, topographical features may have assisted defence more than has hitherto been suspected, eg at London on the north side, the Walbrook stream ponded up against the wall provided a marshy terrain, and further west the river Fleet presented a substantial barrier (see Maloney, this volume).

Bastions are recorded in a wide variety of shapes, with those of hollow construction being greater in number, but solid ones are found at Aldborough, Caistor (Lincs), Mildenhall, and on the east side of London. At Lincoln solid stone platforms set at intervals along the inside of the walls functioned perhaps in the same way as external towers, ie as mountings for rock-throwing artillery. While most bastions appear to be secondary additions on the evidence of straight joints, it should be noted that there are bastions at Caistor (Lincs), Caistor-by-Norwich, Canterbury, Horncastle, and Verulamium (270) which are considered as being bonded into the town walls. In the absence of firm evidence for a late date, the likelihood of the bastions from these towns being contemporary with their respective walled circuits cannot be discounted. On the other hand the late circuits of Cambridge, Catterick, Great Chesterford, and East Stoke apparently have no bastions, and this needs to be explained in the light of the general conclusion that very late town walls were modelled on the Saxon Shore forts. Requiring further thought is the possibility that Cambridge and Great Chesterford could be the southern towns of the eastern defence system just discussed, linking with the bastioned towns of Godmanchester, Water Newton, Great Casterton, and Ancaster, through Lincoln (with its internal turrets) to Brough-on-Humber. Frere sees the development of bastions as 'the result of a single decision (though several years may have elapsed before it was fully implemented)' (Frere 1978, 291). Accordingly, he argues, the latest date should help serve as a pointer to date the rest. At Caerwent evidence for dating the bastion is based on an unworn coin of 330–5 from the early ditch. At Great Casterton the latest date for the addition of bastions comes from two coins of 354 and 358. In London the date of Bastion 6 is bracketed by coins of Constans 341–6 and Theodosius up to 370. However, all this evidence, even when sealed by bastion foundations, points only to a *terminus post quem*, which cannot indicate exactly how late this final stage in Roman defensive architecture occurred; dating a whole system by the latest coin can, as Frere suggests, be misleading. It may be that these improvements in urban defences were the work of Count Theodosius (379–95), for as Ammianus records he helped restore the towns of the province; but it should not automatically be concluded that he directly initiated the construction of bastions (Ammianus Marcellinus, xxviii, 3.2). Quite small towns could have been included in this programme, and it is therefore a vital priority for future studies to establish which of the minor towns remained undefended so that they can be related to the lines of defence proposed in this paper and indicated by bastions.

As with so many aspects of town defences, our understanding of the history of bastions is in its infancy and surprisingly has not yet advanced far beyond Corder's work. There can, however, be little doubt that the building of bastions in the late 4th century immensely increased a town's capacity to hold out against the barbarians, as happened in some instances long into the 5th century. Bastions must have made a significant contribution to the barbarian dread of walled settlements. This success does not support Rivet's view (1958, 96) that 'bastions represent a strategy of despair'. How successful walls and bastions were in Britain is not historically documented, but for Gaul Ammianus records how formidable a physical and psychological barrier they could be to Saxon peoples. Fritigern, leader of the Goths *c* 376, is reported to have said, 'For my part I am at peace with walls' (Ammianus, xxxi, 6.4). Ammianus asserts that some Saxons had a dread of Roman cities or settlements and thought that walls were a trap: 'They avoided these as if they were tombs of their ancestors surrounded by nets' (Ammianus, xvi, 2.12). The small bands of barbarians needed speed and surprise to overcome the problems of getting past the walls into the towns. This is vividly shown at the town of Langres in Gaul when the Emperor Constantius was nearly captured by barbarian raiders but, as Eutropius records, 'he was pulled up inside the walls with ropes after the gates had been closed' (Eutropius, *Breviarium*, IX, 23).

Closely related to the functions of bastions is the question of the *burgi* or defended road settlements, both of which could be related to mobile units of the field army, selected strongholds being used for supply bases and encampments. By the end of the second half of the 4th century these defensible road networks behind the frontiers were vital to Britain and the north-west provinces. On the Rhine and Danube frontiers they may have controlled trade as well as protecting the frontier. In Britain, Rivet and Todd see the *burgi* as part of the 4th century preoccupation with defence, which was expressed in stone fortifications at 'lesser' centres important primarily to the west midlands (Rivet 1975, 112). Webster recognized the beginning of such a chain of *burgi* along Watling Street between Wroxeter and Towcester, which could possibly have extended as far south as Dunstable (Fig 82d; Webster 1971, 38–45; 1975, 53). He believes them to be survivors of a more extensive system protecting a strategic route stretching from Chester to London and even possibly to Dover, though here the evidence is totally lacking. Similarly, an east midlands group has been recognized and enclosures have been found on the Fosse at Brough-on-Fosse (Crococalana), East Stoke or Thorpe by Newark (Ad Pontem), and East Bridgeford (Margidunum), with a solitary example at Ancaster further east between Lincoln and Water Newton (Todd 1975, 211–23). These few recorded examples cannot stand alone and must be part of a more extensive military chain. However, excavation of these small walled road stations has yielded little in the way of structures or finds of military character, though their late date is generally accepted from what little 4th century pottery survives.

At this late period of bastions and *burgi* there was a tendency to narrow or block gateways in town walls, eg the north and south gateways at Caerwent and the south-west and south-east gates at Silchester (Fulford, this volume). However, this may only have been a local part of a general programme for defences and not a response to a specific threat. Similarly, evidence of large-scale rebuilding works has been found at Gloucester and Lincoln in the 4th century, at Verulamium in the 3rd, and at Towcester, where major wall collapses were noted, in the 4th century. Clearly, the dating by unwary excavators of whole circuits from limited investigations which have located such alterations could produce very misleading conclusions. Alternatively, a secure dating for these repairs in the 4th century might suggest a connection with the Theodosian 'restoration' of towns. They could, in addition, be an indication as to whether or not the towns suffered during the barbarian invasion of 367. Evidence of one of the latest rebuilds of a section of wall comes from the London riverside wall at the Tower of London. Here a *terminus post quem* to the repair work was given by a group of coins dated to the early 390s. The original wall was dated at Blackfriars to *c* 360. The date of the rebuild allows for the possibility that it was the work of Stilicho in reorganizing the defences of Britain after 390. It might also just admit the possibility of an early 5th century date, for which the words of Emperor Honorius in 410 'that the cities of Britain should look to their own defences' would be appropriate. Clearly, the survival of towns in Britain and Gaul into the second half of the 5th century, at least as spheres of Roman economic and social influence, must have largely depended on the strength of town walls. It must be significant therefore that in neither country is there much evidence for destruction of these defences; rather they fell through disrepair due to abandonment or neglect.

In summary, this review of the history and development of Romano-British town defences has attempted primarily to show the archaeological detail that has to be understood before historical conclusions can be attempted. Too often in the past working hypotheses have become enthroned as historical fact. Therefore the temptation to create historical horizons must be resisted and the orthodoxy that exists today replaced by interacting models. This flexible approach must be based upon building up a picture town by town, century by century, and only when the data really permit should a broad historical view be attempted.

To this end the following lines of enquiry could be profitable. Firstly, the examination of town defences as an integral part of urban planning, since the walls are strongly related to the boundaries and legal limits of the town. Secondly, the effect of such a major demarcation on property divisions, both within the defences, eg residential and industrial zoning, and outside, eg cemeteries and 'ribbon' development. Thus an understanding of the effect of subsequent growth both intra- and extra-mural can be achieved, which is vital if the true character of a town at any time is to be fully understood. Finally, the quality of dating evidence should be more closely specified and perhaps an 'estimated date of deposition' be introduced to help offset the current emphasis on *terminus post quem* dates.

Note

1 Acknowledgement, however, should be made, as regards London, of the important pioneer work of antiquaries such as Charles Roach Smith working in the 1850s and 60s, followed by Price, Norman, and Reader at the end of the century and in the years immediately before the First World War (see Wheeler 1928, 69–106).

References

Butler, R M, 1959 Late Roman town walls in Gaul, *Archaeol J*, **116**, 25–50

Collingwood, R G, & Richmond, I A, 1969 *The archaeology of Roman Britain*

Corder, P, 1955 The reorganization of the defences of Romano-British towns in the fourth century, *Archaeol J*, **112**, 20–42

Cunliffe, B W, 1973 *The Regni* (Peoples of Roman Britain series)

Frere, S S, 1964a Verulamium—then and now, *Bull Inst Archaeol*, **4**, 61–82

——, 1964b Verulamium, three Roman cities, *Antiquity*, **38**, 103–12

——, 1965 Town defences in Roman Britain, *ibid*, **39**, 137–9

——, 1967 *Britannia: a history of Roman Britain*, 1 edn

——, 1972 *Verulamium*, **1**, Res Rep Soc Antiq London, **28**

——, 1978 *Britannia: a history of Roman Britain*, 2 edn (paperback 1974)

Hartley, B R, 1966 Dating town buildings and structures, in Wacher 1966, 52–9

Jarrett, M G, 1965 Town defences in Roman Britain, *Antiquity*, **39**, 57–9; 139; 227

Johnson, J S, 1973 A group of late Roman city walls in Gallia Belgica, *Britannia*, **4**, 210–23

Ordnance Survey 1978 Ordnance Survey, *Map of Roman Britain*, 4 edn

Rivet, A L F, 1958 *Town and country in Roman Britain*

——, 1964 *Town and country in Roman Britain*, 2 edn

——, 1966 Summing up: some historical aspects of the *civitates* of Roman Britain, in Wacher 1966, 101–13

——, 1975 Summing up: the classification of minor towns and related settlements, in Rodwell & Rowley 1975, 111–14

Rodwell, W, & Rowley, R T (eds), 1975 *The small towns of Roman Britain*, Brit Archaeol Rep, **15**

Todd, M, 1970 The small towns of Roman Britain, *Britannia*, **1**, 114

——, 1975 Margidunum and Ancaster, in Rodwell & Rowley 1975, 211–23

Wacher, J S, 1962 A survey of Roman Britain town defences of the early and middle second century, *Archaeol J*, **119**, 103–13

——, 1965 Town defences in Roman Britain, *Antiquity*, **39**, 225–7

—— (ed), 1966 *The civitas capitals of Roman Britain*

——, 1975a *The towns of Roman Britain*

——, 1975b Village fortifications, in Rodwell & Rowley 1975, 51–2

——, 1978 *Roman Britain*

Webster, G, 1971 A Roman system of fortified posts along Watling Street, Britain, in *Roman frontier studies 1967, Proc 7th Internat Congress* (ed S Applebaum), 38–45, Tel Aviv

——, 1975 Small towns without defences, in Rodwell & Rowley 1975, 53–66

Wheeler, R E M, 1928 *Roman London*, RCHM

Wheeler, R E M, & Wheeler, T V, 1936 *Verulamium, a Belgic and two Roman cities*

X By 1990 it was widely believed that by 411 Britain was de-monetized & without commerce or urban life, towns being used as castles.

The defensive sequence at Silchester

Michael Fulford

Introduction

The Roman settlement of Calleva Atrebatum in the parish of Silchester, Hants, was, as its name implies, a *civitas* capital. The intra-mural area has been extensively excavated, particularly in the period between 1890 and 1909 when most of the gates through the town wall were cleared and planned. (For a comprehensive and up-to-date analysis of the town see Boon 1974.) Otherwise extra-mural defensive ditches were partly traced and the earthwork in Rampier Copse to the south-west of the town was sectioned (St John Hope & Stephenson 1911). A fuller understanding of the defensive sequence had to wait until 1938–9 when Mrs M Cotton carried out an important series of excavations on the line of the wall, the earthen bank immediately behind, and the extra-mural earthworks to the west (Cotton 1947). A further considerable advance in our understanding of the evolution of the settlement's defences was made by Mr G Boon's excavations of 1954–8 which traced the course of the intra-mural ditch referred to as the 'Inner Earthwork' (Boon 1969). Subsequently, as part of the DoE programme of consolidation of the town walls, further excavation took place (1968; 1974–8) along the southern sector including the re-examination of the south and south-east gate (Collis forthcoming; Fulford in preparation).

The defensive sequence (Fig 83)

The Inner Earthwork

The excavations of 1954–8 showed clearly that the Inner Earthwork could be placed early in the development of the settlement. Its course was fairly well defined by aerial survey except to the south-east where it remains obscure. Altogether *c* 32·5 ha (*c* 80 acres) were enclosed. Limited trenching established that the ditch was *c* 13·5 m wide at the top and *c* 3·5 m deep. The precise location of the entrances has not been tested by excavation, but Boon has suggested that the east and west entrances were aligned on the London–West Country road and the south entrance on the Chichester road. It is not possible to be precise about the date of this earthwork. A group of pottery found beneath the denuded bank contained no *sigillata* or imported *terra nigra* or *terra rubra* (Boon 1969, 74–80). The absence of these imported wares and the typology of the coarse pottery pointed to a date post *c* AD 25. Further material from the silts gives a *terminus ante quem* of *c* AD 50/60 for the filling of the ditch (*ibid*, 62–5). Therefore, while this defence could indeed predate AD 43–4 and the advance of Aulus Plautius, Boon preferred a date a little later in the 40s, suggesting that the defence was constructed by the client king Cogidubnus (with official approval) as a counter to the campaigning of Caratacus (*ibid*, 39). Whatever the precise date, the earthwork draws attention to the size of settlement considered worthy of defence at this time.

The Outer Earthworks

Although the Inner Earthwork is the earliest dated earthwork, we should not omit a consideration of the prominent earthworks to the west of the walled area. The stretch to the north-west (Sandy Lands) was sectioned by Mrs Cotton in 1939 but produced no evidence of date (1947, 137–8). The Rampier Copse earthwork was sectioned in 1909 and the description of the pottery found suggests a pre-Roman date (St John Hope & Stephenson 1911, 326). The pottery has not survived, so that we cannot reassess its date. However Mrs Cotton dug a trench in the tail of the rampart in 1939 and recovered some 1st century AD pottery. Both the incompleteness of her section into the rampart and an examination of the stratigraphic position of the pottery make it by no means certain that the pottery does indeed provide a *terminus post quem* for the construction of the bank (1947, 138–40, pl XXXVIII). The surviving drawing of the 1909 section suggests the possibility of two phases in the Rampier Copse bank (Boon 1969, 16–18, pl IX). While as yet undated, the character of the earthwork and the lack of material from beneath it point to an early, possibly pre-Roman date for the original construction of both stretches of standing earthwork to the west of the walled area. Whether these two sections were once linked in order to cut off the promontory on which Calleva stands is not entirely certain. A section dug in 1978 at the southern end of the Sandy Lands earthwork showed evidence of a shallow ditch which suggested incompletion. If these earthworks did originally serve to defend the promontory, their success was only guaranteed as long as the heavy soils surrounding the promontory remained covered in vegetation.

For some time the Outer Earthwork has been considered to have been embodied in a late 1st century AD defensive scheme which, it was believed, enclosed *c* 80 ha (200 acres). Originally this scheme, it was argued, included the ditched annexe to the west of the standing outer earthworks (Primary Outer Earthworks) (Boon 1969, 15–18). Its course was originally traced from the air and later on the ground during the excavations of 1954–8. The ditch was *c* 7 m wide at the lip and *c* 2 m deep. The date of this arrangement is a little uncertain; Claudio-Neronian sherds were found on the remaining ground surface below the presumed site of the bank, but the primary silts also produced later Roman pottery (Boon 1969, 18–21). The relationship of this earthwork with the standing Outer Earthwork to its east has not been demonstrated.

It is difficult to regard this ditch as a defence of this western annexe of the settlement. While aerial survey suggests settlement along the course of the roads out to the west, beyond that ribbon development there is no indication of much settlement nor of organized streets and lanes (*cf* Boon 1974, plan). Indeed, in the northern part of the annexe the plough brought up evidence in 1979 of an extensive 2nd century cremation cemetery. The inner slopes of the Rampier Copse earthwork were also used for cremation burials at about the same time (Karslake 1911). As for the idea that the annexe or primary Outer Earthwork and later the Outer Earthwork proper formed part of a scheme which included land to the east and south of the walled settlement (as marked on Boon 1969, pl I; 1974, plan), trial excavation in 1978 could find no support.

Excavations in 1980 added a new dimension to the question of late 1st century defences with the discovery of a probable ditch within the later walled area. It is located some 22 m west of the east gate (see below). The original width may have been *c* 10 m and the deepest excavated point was 4·2 m below the present ground surface. The ditch appears to follow a NNW–SSE course. The lowest

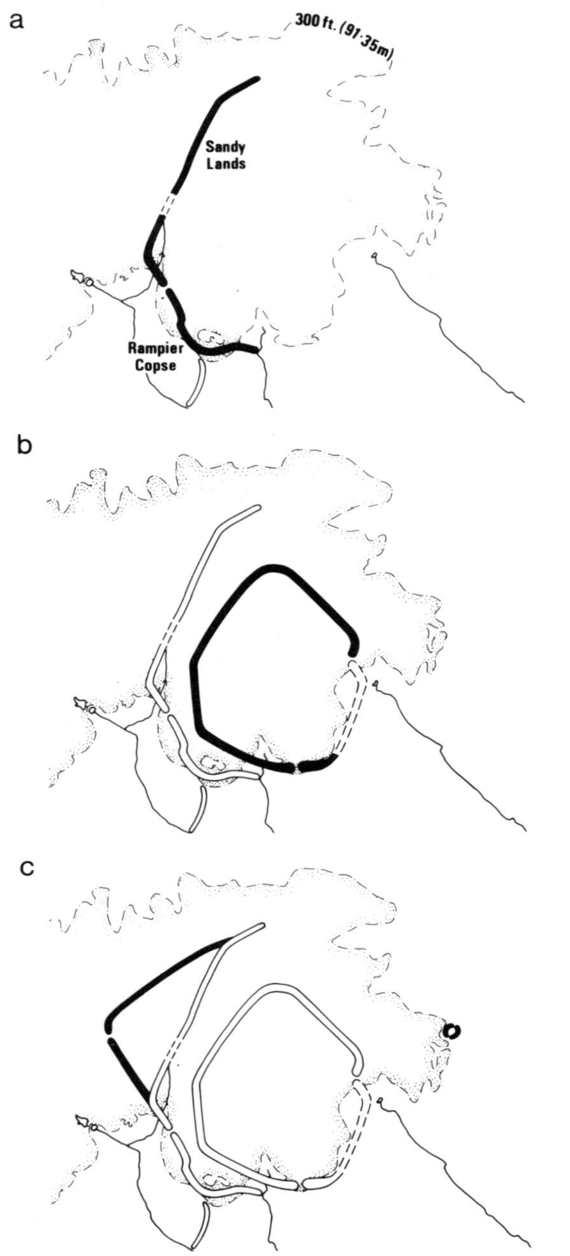

Fig 83 Silchester: outline of the defensive sequence: A pre-Roman; B mid 1st century (Inner Earthwork); C ?later 1st century (Outer Earthwork annexe); D late 2nd century earthen rampart and mid-late 3rd century town wall

silts contained early Flavian samian. This recalls the early Flavian defensive ditch recently discovered in Winchester (Biddle 1965, 235–8, pl LXIX; 1970, 281–5). At the beginning of the 2nd century a trench for a palisade or fence was dug along the southern lip of the ditch. The layers that eventually sealed this contained mid to late 2nd century pottery. Thus the ditch was out of use at about the time when the town rampart was constructed (see below).

2nd century defence (Figs 84, 85)

Thus the phase of defensive work which followed the Inner Earthwork was the construction of the earthen rampart whose course was later followed by the town wall. The area enclosed (*c* 40 ha; 100 acres) is comparable to the area within the Inner Earthwork except for the addition of the 'temple precinct' on the eastern side. This rampart has been sectioned to the subsoil in ten different places and its construction has been shown to be uniform. It is of dump construction and at present stands to a height of *c* 3·5 m above the Roman ground surface (see eg Cotton 1947, figs 1–3). The materials used include gravel and sandy clay derived from the digging of the double ditches fronting it. So far there is no evidence of any wooden structure either within the bank or standing upon it as a tower. At the south-east gate the front appears to have been revetted with nothing more substantial than a wattle fence. Similarly at the south-west corner there was no trace of revetment; the front of the bank appeared to slope straight down into the inner ditch.

Contemporary with the rampart and, if anything, technically predating it are the main gates (Fig 84). Their position and relative alignment have long suggested that they may have been earlier than the walls. Excavation at the south gate in 1975 showed clearly that the foundations of the gate were present when the bank was thrown up. Similarity in plan with the north gate points to the same sequence there. The main east and west gates are also similar in plan and slightly askew to the course of the wall and their construction is dissimilar to that of the city wall in that the west gate is built of ironstone alternating with bonding-courses of tile. Internal walls are constructed of flint and tile in alternate double-bonds (Fox & St John Hope 1890, 755). Although the east gate is less well preserved, excavation in 1908 showed that it abutted the town wall with a straight joint on the northern side. Once again tile was used in the footings, quoins, and bonding-courses (St John Hope 1909, 474–6, pl 84). Use of tile was also a feature of the south gate. Thus each pair of main gates betrays similarities in plan as well as in materials used. All the gates described are clearly distinguished from the town wall in the way that tile, rather than stone, formed an integral part of their construction. It is clear that all four main gates and at least one minor gate (that at the south-east) are earlier than the 3rd century town wall. Given the relationship between the earthen rampart and the south and south-east gates, it also seems reasonable to assume that all were designed to accompany that earthwork.

The latest dated material from beneath it includes samian and BB1 of late Antonine date (*c* AD 160–80) (Cotton 1947, 129–30; Fulford in preparation). It is interesting to note that material of this date seems to derive from cultivated soils rather than structures. The latest date for structural evidence or occupation beneath the bank is the early 2nd century (Trajanic/Hadrianic) (Cotton 1947, 127; Fulford in preparation).

So when considering the positioning of the defence it does not seem that currently occupied sites were built over. Indeed with the environmental evidence from the

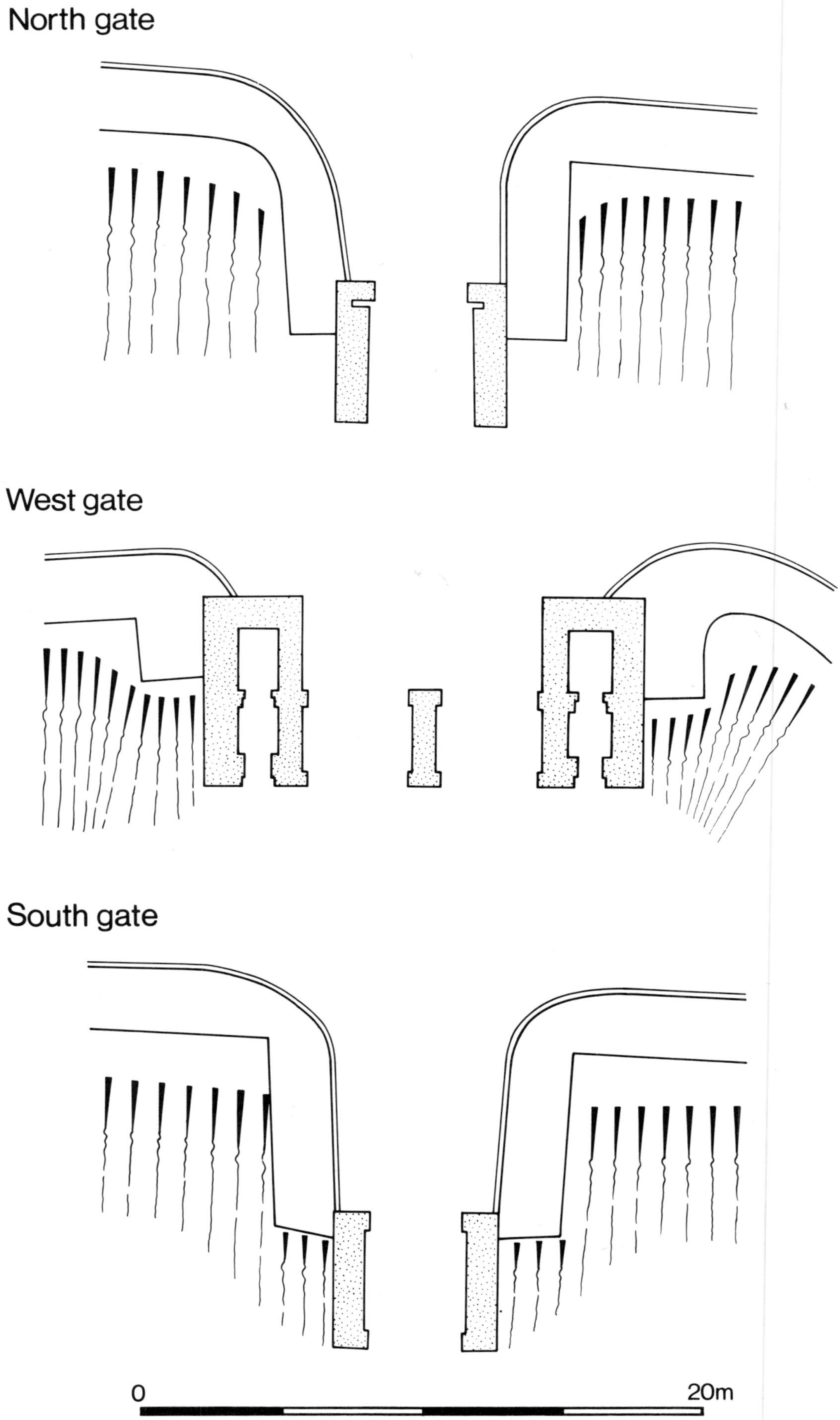

Fig 84 The north, west, and south gates

south-east gate suggesting the presence of nearby waste land it would seem that the area enclosed was generous. There are obvious exclusions, such as the ribbon development along the main east-west road, the amphitheatre, and some isolated buildings identified from the air, but none that could be practicably enclosed. Altogether the bank takes in the majority of the well drained plateau-gravels which form the promontory upon which the settlement lies, and not the lower lying clay soils of the Wickham and Titchfield series.

Despite the generous amount of land enclosed, Mr D W A Startin's calculations on the time necessary to construct the rampart indicate that the work need not have taken long. He estimates that 300,000 man-hours were required, ie 300 men working 100 days of 10 hours. The task could have been accomplished within one year, probably between spring sowing and harvest time. Even if we take account of the gates, the defence could have been finished to answer an emergency in the troubled times of the late 2nd century in Britain.

The town wall (Figs 85, 86)

The last main phase in the defensive sequence at Calleva was the construction of the city wall (Boon 1974, 100–7). This involved cutting back the front of the earthen rampart and building a wall *c* 3 m thick, reduced to *c* 2·3 m in width at a height of 2·8 m, except where the counterforts carried the original thickness to full height. The highest surviving stretch of wall stands to *c* 4·5 m above Roman ground levels. It is built in horizontal stages of *c* 1 m height, each containing 4–5 courses of flint and capped with 1–2 courses of bonding slabs (each *c* 0·75–1 m long). The outer face was finished with a hammer-dressing. It is important to stress that unlike the gates and many of the buildings within the walls tile-coursing was not used. Reused monumental masonry has also not been found. The wall was fronted by at least one wide, flat-bottomed ditch and in some places by two.

The most abundant material used is flint for which the nearest source is *c* 15 km away. The lithology of the bonding slabs indicates a variety of sources, most of which are some distance from Calleva. Dr B W Sellwood's preliminary findings suggest a more complicated situation than was originally thought from the samples submitted to Melville (1947). Coral ragstone (Corallian—Upper Jurassic) and coarse pisolite are used in roughly equal measures with a coarser glauconitic and pebbly shelly quartzite (Upper or Lower Greensand). There is some ironstone (Tertiary) which could have been obtained locally from the plateau gravels. A number of sources are clearly involved, but we can exclude the quarries which provided the best quality limestone from the deep Jurassic (such as Bath). The likely origin of the limestone includes the area around Faringdon in Oxfordshire and elsewhere on the eastern margins of the Jurassic. This stone is available at a distance of 50–80 km from Calleva. Greensand could be obtained from nearer sources. It is interesting to speculate whether this material derived from quarries within the *civitas* (cf Boon 1974, 101), and it is worth pointing out that the limestone employed in Bath and Cirencester are different. This might suggest that Calleva did not obtain its material from quarries in either the *civitas* of the Belgae or that of the Dobunni.

It is difficult to be precise about date. The fill of the wall behind the construction trench contained some Oxfordshire red colour-coated sherds and the later forms of BB1. While we cannot be certain about the starting date of these types, present evidence points to a date post *c* 225.

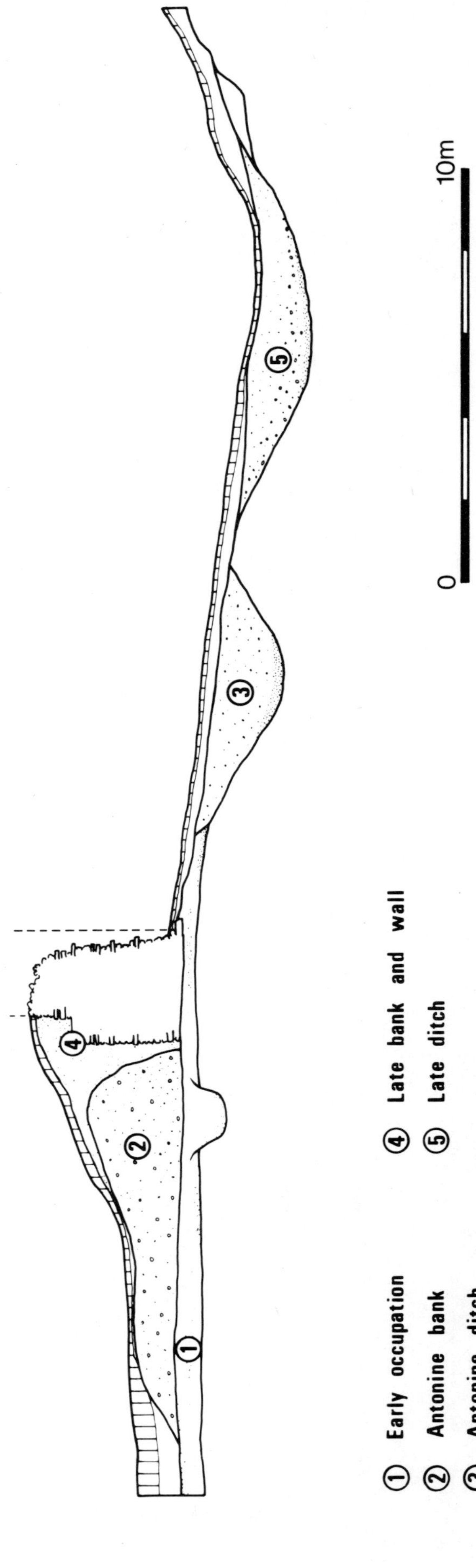

Fig 85 Reconstruction of the 2nd–3rd century defensive sequence

Fig 86 The 3rd century town wall and the south gate

It is likely that the wall was constructed between *c* 225 and *c* 275, perhaps between *c* 250 and *c* 275. If the wall was later than *c* 275 one might have expected bastions to be present, on the model of the Saxon Shore forts.

The construction of the wall clearly took considerably longer than the building of the rampart and, given the character and diversity of the material, the work involved more extensive use, both human and material, of the resources of the *civitas*.

Later Roman alterations

There was no substantial addition or alteration to the defences in the late Roman period. Bastions, for example, do not appear to have been added. The blocking of gates can be clearly seen at the south-east postern and possibly also in one carriageway of the west gate. There the excavators reported finding 'masses of uncut carstone' (probably bonding stone from the wall) with small fragments of architectural masonry. It is not clear whether this stone was mortared together and the only visual record of it appears to be the contemporary model of that gate. There is no evidence of date for the dumping of this material and it might well have occurred during the collapse or robbing of the wall and gatehouses. At the south gate there was no convincing evidence of blocking although rubble had been strewn outside the gate after *c* AD 350–60. The presence of this material suggests gradual disuse rather than blocking.

A final comment should be made about the Grim's Bank to the north-west of Calleva. In the past this has been interpreted as a late or sub-Roman boundary of the *territorium* of the town. No dating evidence has so far been recovered from it and a prehistoric date cannot be ruled out (Astill forthcoming). The relationship between this earthwork and the Silchester–Dorchester road needs further investigation.

References

Astill, G G, forthcoming Excavation at Grim's Bank, *Berkshire Archaeol J*, **70**

Biddle, M, 1965 Excavation at Winchester 1964, third interim report, *Antiq J*, **45**, 230–64

———, 1970 Excavation at Winchester 1969, eighth interim report, *ibid*, **50**, 277–326

Boon, G C, 1969 Belgic and Roman Silchester: the excavation of 1954–8 with an excursus on the early history of Calleva, *Archaeologia*, **102**, 1–82

———, 1974 *Silchester: the Roman town of Calleva*

Colis, J, forthcoming Excavation at Silchester 1968

Cotton, M A, 1947 Excavations at Silchester 1938–9, *Archaeologia*, **92**, 121–67

Fox, G E, & St John Hope, W H, 1890 Excavation on the site of the Roman city at Silchester, Hants, *ibid*, **52**, 733–58

Fulford, M G, in preparation Excavation at Silchester 1974–8

Karslake, J B P, 1911 Note on discoveries in the outer entrenchment, in St John Hope & Stephenson 1911, 330–2

Melville, R V, 1947 Building stones, in Cotton 1947, 143

St John Hope, W H, 1909 Excavation on the site of the Roman city at Silchester, Hants, in 1908, *Archaeologia*, **61**, 473–86

St John Hope, W H, & Stephenson, M, 1911 Excavation about the site of the Roman city at Silchester, Hants in 1909, *ibid*, **62**, 317–32

In the introductory paper to this section (pp 78–84 above), Brian Hobley has attempted to set both the development of urban defences and the history of their study into a wider context. The purpose of the present paper is to describe and discuss the evidence for the defensive sequence at one type of urban centre in Britain—that of the *colonia*, the highest status of provincial town. It also seeks to examine whether the *coloniae* formed a group in terms of their fortifications, in what ways they differed from other towns, and the significance and implications of any differences.

The 1st century colonies at Colchester, Lincoln, and Gloucester were founded as settlements for legionary veterans (Frere 1974, 231, 273–4; Wacher 1975, 37–8). The creation of three *coloniae* on the sites of earlier fortresses was a useful expedient whereby time-served legionaries could be discharged, form a military reserve, and receive their due grant of land with minimal disturbance of the native population. In contrast, the sites of the fortresses at Exeter (Bidwell 1979) and Wroxeter (Webster 1980) became tribal centres. The only other known *colonia* in Britain, at York, is generally believed to have been promoted to this status in the early 3rd century (Wacher 1975, 156).

The foundation of the colony at Colchester took place in AD 49. Tacitus tells us that it had a dual purpose: as a military base in the hinterland of the frontier zone, and as a model of Roman urban life (*Ann*, 12.32). This statement has some support from archaeological excavations: some military barracks were not demolished but continued to be used in modified form in the early *colonia*, while at the same time public buildings were being erected and a new street system laid out to the east of the fortress site (Crummy 1977, 76–81). On the other hand, the early colony had no defences: after the events of AD 61 (Webster 1978, 113–20; Carroll 1979), this mistake was not repeated.

Evidence has been found near the Balkerne Gate, the later west gate of the town, for a defensive system built soon after the Boudican revolt, consisting of a ditch and presumably also an earth bank (Crummy 1977, 95–100). This line was abandoned *c* 75 when the defended area was apparently extended westwards, the former western boundary being marked by a monumental arch. The arch was incorporated into the new Balkerne Gate when the city wall was built along the line of the earliest western defences in the early 2nd century.

Some fine stretches of the wall at Colchester survive (Fig 87), although in places the front was refaced in the medieval period. Built of alternate layers of *septaria* and mortar, with tile courses in both inner and outer faces, it was about 3 m wide at its base and slightly narrower above. Several interval towers have been found, all presumably contemporary with the wall. Apart from the blocking and refacing of the Balkerne Gate (late Saxon–early Norman?), there are only slight traces of rebuilding before the late medieval refacing. The dating evidence shows that the wall was built between *c* AD 100 and *c* AD 150.

A major problem here is whether or not the building of the earthen rampart was from the start seen as an integral part of the scheme. The latest pottery in the bank as excavated in 1972 at Lion Walk was no later than *c* 150 (Crummy 1977, 91–2, with figs 15, 16). Elsewhere the rampart was found to contain Colchester colour-coated pottery once thought to be of late 2nd century date, but now dated to before 150. Beneath the bank and overlying the wall's construction levels there were clear remains in places of several road surfaces and layers of occupation material. It seems unlikely that all of these could have been associated with the building of the wall, even though it was nearly 2500 m long and probably took several years to complete. The 3 m wide wall could easily have accommodated a wall-walk, although wide free-standing walls were most unusual in Britain before the late 3rd century. Their erection at Colchester at such an early date in comparison with other British towns must surely have been connected with the city's colonial status.

The two late 1st century colonies at Gloucester and Lincoln, like the new tribal capital at Exeter (Bidwell 1980, 46–7), initially made use of the earlier legionary defences. At both sites a wall was erected in the early 2nd century, and this constituted a refacing, rather than a rebuilding, of the legionary defences. The sequence of the so-called upper *colonia* at Lincoln is now fairly clear, following extensive research which has recently been published (Jones 1980).

In all, 22 sites on the defences of the upper *colonia* at Lincoln have produced evidence of one sort or another. The most valuable results have come from excavations, but an attempt to reconcile the conflicting evidence from various sites has shown how inaccurate and inconclusive single trenches can prove. The evidence of extant fragments of the walls and of antiquarian reference, can, in conjunction with dated contexts from excavation, fill out and refine the picture considerably.

The study of the surviving remains of the city wall has involved the preparation of detailed drawings of selected, representative examples of both face- and core-work and more general longitudinal elevations to set these into context. In general, the core was formed of coursed limestone rubble with an abundant amount of mortar (*opus caementicium*) (Fig 88), but those fragments which are known to be of late date appear to have been more clearly layered, at least in their lowest courses. The stone types were

Fig 87 Colchester: city wall immediately south of Balkerne Gate (visible extreme left), Roman foundation level same as modern ground level. Upper rough coursing is post-Roman (photo: Alison Colchester)

Fig 88 Lincoln, East Bight: rear 'face' of extant fragment of rebuilt wall. Note regular layering of courses beneath overhang (photo: H N Hawley)

analysed by taking samples from each of at least one hundred stones from both core and facing at various sites. The stone employed was nearly all derived from the lower Lincolnshire limestone, obtainable very close to the city. This provided the better quality stone for building, oolitic limestone, as well as non-oolitic limestones.[1] Most parts of the wall examined, apparently those which had been rebuilt, contained about 80% oolitic limestone, but the narrow wall built in the early 2nd century had a much smaller proportion. This suggested that different quarries were being used, and that the better beds were not used until well into the 2nd century. Examination of the legionary tombstones, however, showed that only oolitic stone was used for monumental work from an early date.

A similar pattern emerged from the study of mortars, where three different types were distinguished. The sand and gravel used in the early wall apparently came from a source to the south or to the west of the city, different from that used later, whose source was the beds to the south-west of the city (Jones 1980, fig 48). Study of the sources of building materials in this way, not only from defensive walls, of course, can help to build up information on this neglected aspect of the economy of Roman Britain. Toolmarks and stone sizes were also examined, with less conclusive results, at least from an historical perspective.

The Neronian legionary rampart at Lincoln consisted of a bank of clay and sand revetted front and back by timbers, as first found by Graham Webster in the 1940s (Webster 1949). The interval towers were given new projecting fronts in the early Flavian period, and the existence of the projections (unusual at this time) probably contributed to the decision to build the new wall clear of the early rampart, rather than to cut the rampart back for its insertion.

Excavations by both Thompson and Whitwell on the northern and eastern defences (Thompson 1956; Thompson & Whitwell 1973; Whitwell in Jones 1980) strongly suggested a *terminus post quem* for the construction of the wall of *c* AD 100. The dating material was obtained from beneath the wall and from the material between the wall and the front of the legionary rampart. Similar contexts investigated by Petch, however, on the southern defences (Petch 1960) appeared to indicate an Antonine date, while his work south of the East Gate in 1959 produced Hadrianic material predating the wall.

Re-examination of both the stratigraphical evidence and the pottery (Jones 1980; Darling forthcoming) seems to show that the wall was in fact built after 120, but that the Antonine material could relate to a rebuilding phase contemporary with the erection of the *colonia* rampart behind the wall not before the end of the 2nd century.

The wall (Fig 89) was about 1·2 m wide, and was founded on the rammed fill of the first legionary ditch. Phallic carvings found at two places on the early wall, as at sites on Hadrian's Wall, are not necessarily evidence of masons trained in the army, but it is difficult to know how much any military influence was due to retired specialists rather than to direct military work. A possible rampart-walk was noted by Thompson in 1953 at a point, allowing for later compaction, about 3 m above the Roman ground level. The top of the wall was probably about 1·5 m higher. Apart from the construction of the walkway, the legionary rampart was little altered until the end of the century, while the gates were merely clad in stone until the 3rd century.

The early sequence that has emerged over the last twelve years or so at Gloucester is similar (Hurst 1972, 28–32; Heighway forthcoming). Again the army may have been involved in the construction: the first colonists appear to have lived in barrack-like buildings (Hurst 1972, 39–40). The wall built soon after the colony's creation,

Fig 89 Lincoln: front of early narrow wall, excavated at East Bight. Bottom course offset, sitting on rammed fill of legionary ditch.

Fig 90 Gloucester: foundations of the North Gate built c *AD 100, looking south-east (photo: Gloucester City Museum)*

much obscured by the walls of later periods, had a foundation only 1 m wide, but, in places at least, was apparently wider above this. No substantial additions to the rampart were visible: probably the remnants of the legionary bank were used. As at Lincoln, the wall sat over a legionary ditch, but unlike at Lincoln the north and east gates appear to have been rebuilt completely in stone at this time (Fig 90).

The refurbishment of the fortifications at Lincoln and Gloucester in the early 2nd century has been seen as an aspect of Trajan's policy of consolidation of the province, which is also reflected in the erection of similar stone defensive walls at the three legionary fortresses of Caerleon, Chester, and York, and at the new fort in London (Frere 1974, 146–7). This would remind us again that the two *coloniae*, initially at least, had a quasi-military purpose—to act as a reserve behind the frontier zone. Unfortunately, the dating evidence for the walls at Lincoln and Gloucester is too vague for us to know whether rebuilding work began as early as the Trajanic period, so that the decision to provide stone walls could alternatively have been part of Hadrian's policy for all three colonies.

In any case, the granting of imperial permission at this early date, possibly in response to requests from the new communities, marks them as receiving preferential treatment: no other towns in the province were allowed walls until the following century. The two military *coloniae* in Germany, at Cologne and Xanten, were given walled fortifications soon after their foundation (see the papers by Hellenkemper and Precht, this volume). These were necessary protection, in view of their situation on the Rhine frontier.

From the mid/late 2nd century, however, the development of the defences of the British *coloniae* did show more affinities with that of the non-colonial towns. The evidence for the Colchester wall and rampart has been summarized above. At both Lincoln and Gloucester, a system of stone towers was added on the inside of the wall in the mid or late 2nd century. At Lincoln, at least one gate was rebuilt on a monumental scale in the 3rd century (Thompson & Whitwell 1973, 140–50). The ornamental element reflecting civic pride, which was apparent in the Balkerne Gate at Colchester, is visible again at Lincoln's East and North Gates. At roughly the same time the Lincoln rampart was widened to about 10 m and its heightening may have accompanied a heightening of the wall. The rampart at Gloucester was also heightened in this period, when many of the peregrine towns of Britannia were being provided with new defences (Frere 1974, 283–6). We cannot be very confident about the precise dating, however, still less about its historical context.

The end of the 2nd century also saw the construction of the first fortifications around the lower *colonia* at Lincoln. These were found to consist of a wall 1·5 m wide with a contemporary rampart at Motherby Hill and The Park on the western side (Colyer 1975, 231–5, 248–50), but apparently with a timber palisade on the crest of the bank at Silver Street on the eastern side (Wacher in Colyer & Jones 1979, 81–3). The apparent discrepancy is not as yet understood and the sequence of construction may have varied around the circuit in view of the time needed to build the wall. It is worth pointing out, however, that the tips of rampart material at Silver Street appeared to slope towards a feature (the early wall?) east of the palisade. Towers were added to the lower circuit during the 3rd century.

It was probably in the 3rd century that the *colonia* at York was first defended. Unfortunately, little is yet known of these defences, which have been noted on the west side but have not been systematically investigated (RCHM 1962, 49).

By the middle of the 3rd century then, there was little difference between the defensive systems of the *coloniae* and those of the other major cities of the province. This reflects in some ways the blurring of differences in legal status and administrative functions which had taken place.

From the late 3rd century the distinction between civil and military was also becoming hazy. The fortresses which had preceded the *coloniae* had been built on navigable rivers, and Lincoln and Colchester in particular were now exposed to raids from across the North Sea (Simmons 1980). We must further bear in mind that in the 4th century Lincoln was probably a provincial capital, probably also a bishopric, and, it has been suggested, a military headquarters for the field army (Johnson 1980, 5, 9; Mann 1961; 1977, 14). York was of no less importance. In reviewing the evidence for late defences, I omit questions of weaponry and tactics, and of the possibility of a town militia (see the papers by Baatz and Webster, this volume).

At York, the evidence is again vague. The fortress's defences were rebuilt in the early 4th century, and we might expect contemporary work on the *colonia*. The only evidence for late work at Colchester is a layer dating no earlier than the mid 3rd century which could be interpreted as rampart-heightening (Crummy 1977, fig 16, layer 21), and the widening of the ditch, which now ran across the carriageways outside the Balkerne Gate. The wall may have been heightened at the same time as the new work on the rampart was undertaken.

At Gloucester there is some evidence for rebuilding of the wall in the 3rd century—the so-called 'Period 4', which was stratigraphically later than the late 2nd century rampart and earlier than the later rebuild. The 4th century work here (Period 5) involved rebuilding of the walls, whose foundations at least were up to 4·5 m thick, and the addition of external towers. These alterations are now thought to have taken place before the middle of the century, probably around AD 320–40. The addition of external towers and the digging of a wide U-shaped ditch were either contemporary or later operations.

Fourth century thickening of the wall at Lincoln is known at about ten points on the upper circuit, and at most of the sites investigated on the lower enclosure (Jones 1980). Unfortunately there is little precise dating evidence for the rebuilding, and we cannot be sure if the alterations were all part of one scheme. Pottery dating suggests that the upper defences were refurbished no later than the early 4th century. In the lower enclosure the dating is again vague (Colyer 1975, 241–3, 255–7).[2]

The wall was strengthened on each side of the North Gate, and similarly on each side of the two interval towers known on the north side (Figs 88 & 91). Perhaps only small lengths close to the towers were left unthickened. The highest extant fragment of the late wall, north of the East Gate, stands *c* 6 m above the Roman ground level. This is comparable to the recent discoveries at Canterbury and London (Maloney, this volume), and cannot have been far from its full height. The widths recorded for the late wall are between *c* 2·5 m and *c* 4·5 m, the usual range for late walls in the Western Empire (see Johnson, this volume). The wide discrepancies in width are not yet fully explained. At two points, structures have come to light which constituted more than wall thickening; rather they were solid platforms, which appeared to make the adjacent internal towers redundant. If our available evidence did not argue against it (Marsden 1969, 195–8; see also papers by Baatz, Hassall, and Todd, this volume), it would be tempting to suggest that the platforms, if not the wider

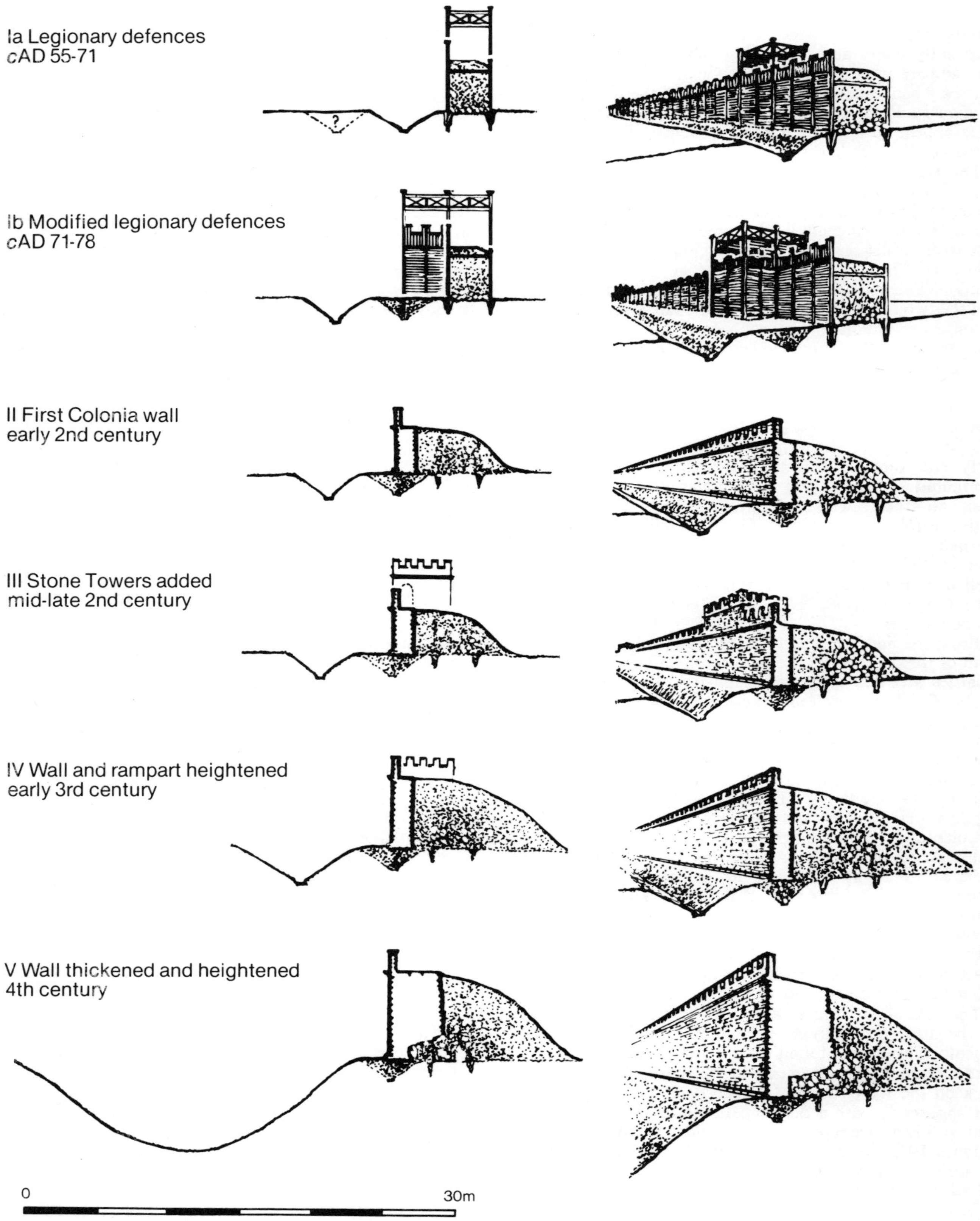

Fig 91 Reconstruction drawing of the sequence for the upper enclosure at Lincoln

wall, were designed to carry artillery machines. There is no evidence as yet from Lincoln, except at the gates, for the externally projecting towers which were common in Britain from the mid/late 4th century. The rampart of the lower town was enlarged in some places; that either side of the gateway on the western defences of the lower town is especially noteworthy, extending about 25 m back from the rebuilt wall. The coin series in the road surfaces through the gate goes on until Arcadius (Colyer 1975, 241–3).

As at Gloucester and Colchester, there was a single wide ditch at Lincoln, here about 25 m wide. The metalled berm found next to the towers of the lower West Gate was covered by a rubbish deposit containing coins and pottery of late 4th century date (Darling 1977).

In the late Roman period then, each of the *coloniae*, like the other cities of the province, was provided with a strong wall backed by a large bank and fronted by a wide ditch. Colchester had some of these elements from the early 2nd century; their construction at Gloucester and Lincoln may have been as late as the mid 4th century. The formidable barrier thus created did not and could not save the Roman economic system, but acted as a shell in the succeeding centuries and influenced the layout of the towns through to the present day.

Acknowledgements

The production of this paper would not have been possible without the cooperation of my colleagues in Colchester and Gloucester, respectively Philip Crummy and Carolyn Heighway. Both generously gave of their time in conducting me around the visible remains of the defensive circuits, and both have kindly given further help during the preparation of the paper for publication, as well as providing photographs. Professor S S Frere also read and commented on an earlier draft. Maggi Darling advised on the pottery dating of the Lincoln defences. Brian Hobley allowed me to see a draft of his own contribution before publication. Greta Exton and Audrey Morton typed earlier versions, and Elizabeth Nurser edited the copy.

Notes

1 For a discussion of the geological aspects, see M Fenton in Jones 1980, Part C.

2 The late rampart levels at the Park did, however, produce a small sherd of a Mediterranean amphora. There is a possibility that this could be of 3rd/early 4th century date, but it is more likely that it postdates *c* AD 370. On grounds of historical probability I have favoured the earlier dating, but further research could involve a radical rethink. I am grateful to my colleagues Lauren Adams and Maggi Darling for discussing the sherd with me.

References

Bidwell, P T, 1979 *The legionary fortress and basilica and forum at Exeter*

——, 1980 *Roman Exeter: fortress and town*

Carroll, K K, 1979 The date of Boudicca's revolt, *Britannia*, **10**, 197–202

Colyer, C, 1975 Excavations at Lincoln 1970–72: the western defences of the lower town: an interim report, *Antiq J*, **55**, 227–66

Colyer, C, & Jones, M J, 1979 Excavations at Lincoln, second interim report: excavations in the lower town 1972–78, *ibid*, **59**, 50–91

Crummy, P, 1977 Colchester, fortress and *colonia*, *Britannia*, **8**, 65–106

Darling, M J, 1977 *A group of late Roman pottery from Lincoln, The archaeology of Lincoln*, **16**/1

——, forthcoming *Roman pottery from the upper defences, The archaeology of Lincoln*, **16**/2

Frere, S S, 1974 *Britannia: a history of Roman Britain*, 2 edn

Heighway, C, forthcoming *The north and east gates of Colchester*

Hurst, H R, 1972 Excavations at Gloucester 1968–71: first interim report, *Antiq J*, **52**, 24–69

Johnson, J S, 1980 *Later Roman Britain*

Jones, M J, 1980 *The defences of the upper Roman enclosure, The archaeology of Lincoln*, 7/1

Mann, J C, 1961 The administration of Roman Britain, *Antiquity*, **35**, 316–20

——, 1977 *Duces* and *Comites* in the 4th century, in *The Saxon Shore* (ed D E Johnston), CBA Res Rep, **18**, 11–15

Marsden, E W, 1969 *Greek and Roman artillery: historical development*

Petch, D F, 1960 Excavations at Lincoln 1955–8, *Archaeol J*, **117**, 40–70

RCHM 1962 *An inventory of the historical monuments in the city of York*, **1**, *Eburacum: Roman York*, Royal Commission on Historic Monuments

Simmons, B B, 1980 Iron Age and Roman coasts around the Wash, in *Archaeology and coastal change* (ed F H Thompson), Soc Antiquaries London

Thompson, F H, 1956 Roman Lincoln 1953, *J Roman Stud*, **46**, 22–32

Thompson, F H, & Whitwell, J B, 1973 The gates of Roman Lincoln, *Archaeologia*, **104**, 129–207

Wacher, J S, 1975 *The towns of Roman Britain*

Webster, G A, 1949 The legionary fortress at Lincoln, *J Roman Stud*, **39**, 57–78

——, 1978 *Boudica: the British revolt against Rome AD 60*

——, 1980 A note on new discoveries at Viroconium (Wroxeter) which may have a bearing on Hadrian's frontier policy in Britain, in *Roman frontier studies 1979* (eds W S Hanson & L J F Keppie), Brit Archaeol Rep, **S71**, 291–6

Introduction

This paper will describe the results of recent archaeological excavations and research which have contributed to knowledge of the Roman defences of London. Although the text will refer to earlier records as necessary, anyone wishing to attempt a more comprehensive study will certainly need to consult RCHM 1928, 69–106; Merrifield 1965, 48–9, 68–72, 101–13, gazetteer 298–325; Grimes 1968, 47–71; and Marsden 1965, 135, fig 1; 1969, 20–3; 1970, 2–9.

The landward wall constructed *c* AD 200, about 150 years after the foundation of London, is the first known defensive circuit of the settlement. However, there were fortifications in the area before this date—one of which influenced the course of the landward wall—and these will be briefly mentioned below, together with relevant events.

That London was founded after the Roman invasion of AD 43 has never been seriously in doubt, but it has been argued that the first settlement did not develop before AD 50 (Marsden 1980, 17–29). The focal point was the bridge (which may have been located recently: Milne 1982) and an early fort defending the bridgehead on the north bank of the Thames has been conjectured (Merrifield & Sheldon 1974, 188–91). There is evidence of a temporary military fort of early date on the north-eastern outskirts at Aldgate (Chapman 1973) and the possibility of another on the west (Grimes 1968, 183–4). Nevertheless, the assumption that London had a military origin has been challenged (Marsden 1980, 11–29). What Tacitus makes clear is that by AD 60 the settlement was flourishing, 'a place not indeed distinguished by the title of *colonia*, but crowded with traders and a great centre of commerce', and he perhaps implies that there were no defences (*Ann*, 14.33, Oxford edn).[1] In that same year, the first settlement was destroyed during the Boudican revolt. After considerable growth during the Flavian period, by the early years of the 2nd century London seems to have achieved the status of provincial capital (Marsden 1980, 79–96). It was not long after that a small stone fort of standard military design was constructed at Cripplegate on the slightly higher ground at the north-western edge of the urban area (Grimes 1968, 15–40). The fort covered nearly five ha and could have housed about 1500 men. In the circumstances it seems likely that the primary function of the fort was to accommodate the Governor's staff and guards (as well as soldiers in transit), rather than to defend London from attack (Merrifield 1965, 44–5; Marsden 1980, 80–3). Although it was to be another 75 years or so before London was protected by a stone wall, earlier events were to play a part in determining its eventual course.

The landward wall

Course

Amongst the considerations influencing the course of the wall would have been the extent of the settlement and possibly the established position of the urban boundary. The formal limits of Roman London at a particular time should be reflected by the position of the contemporary cemeteries (RCHM 1928, 29–32, 153–63), since it was the Roman custom, enforced by law, to bury the dead outside

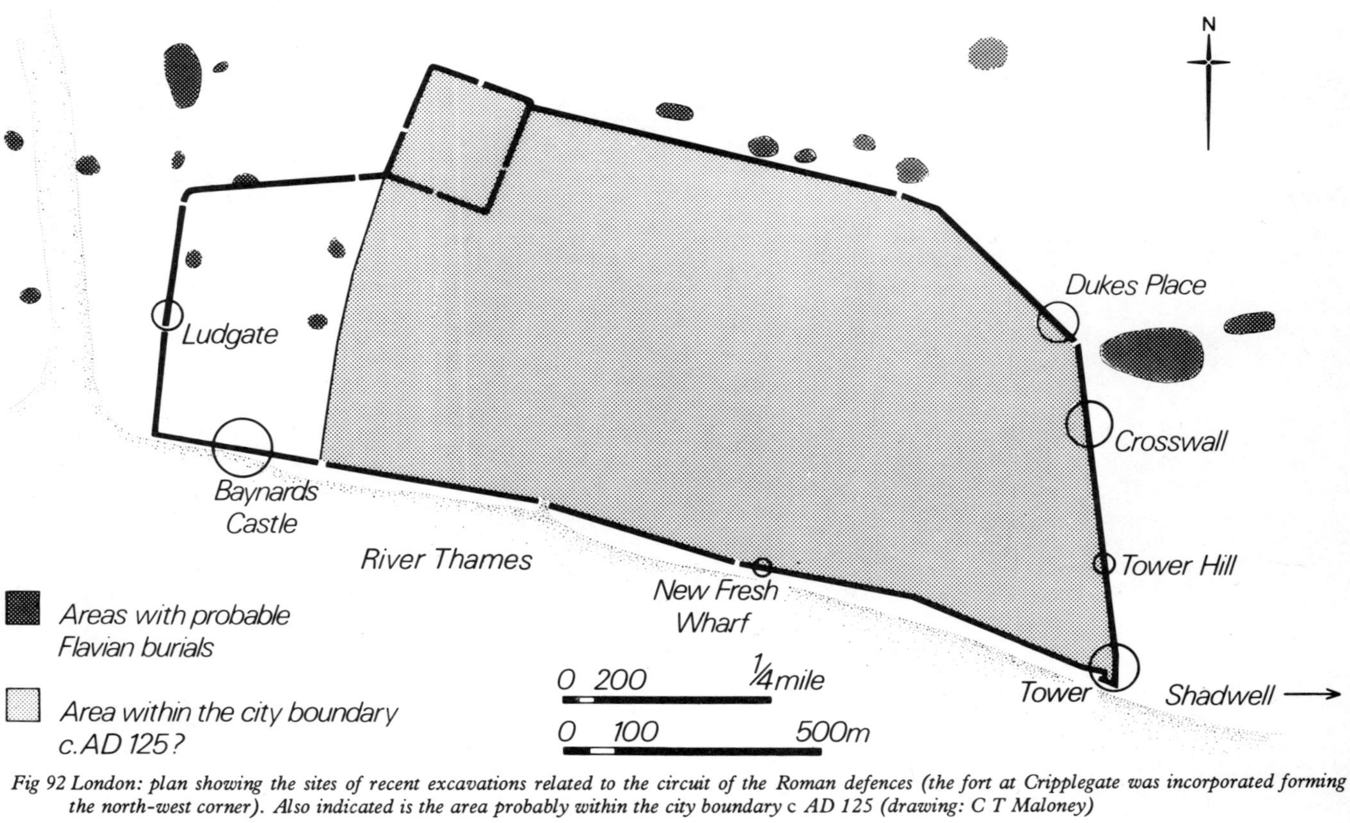

Fig 92 London: plan showing the sites of recent excavations related to the circuit of the Roman defences (the fort at Cripplegate was incorporated forming the north-west corner). Also indicated is the area probably within the city boundary c *AD 125 (drawing: C T Maloney)*

the inhabited area. On the west side of the settlement, within an area later bounded by the wall, the presence of cremations ranging in date from pre-Flavian to at least the early 2nd century indicate that this area lay outside the urban limits during this period (Fig 92).[2] Thus when the fort was built, the boundary on this side of the settlement may have proceeded south from its south-west corner (Marsden 1976, 47–9). If an early circuit of earthwork defences existed it is perhaps along this line that it might most easily be discovered. In a recent study of the western cemeteries it was noted that cremations from the mid 2nd to early 3rd century lie (with one exception) outside the line of the wall although at least some of them must antedate it (Bentley forthcoming). This may imply that during the first half of the 2nd century a new western boundary was established which was later followed by the wall. The eastern cemeteries south of Aldgate contain burials ranging in date from the 1st century to the 4th, and evidently lay wholly outside the course adopted by the wall (RCHM 1928, pl 55). In the northern cemeteries most of the burials have been recorded on extra-mural sites and cover a period from the first half of the 1st century probably into the 4th century. However, several cremation urns together with a coin of Antoninus Pius were found just inside the wall at Bishopsgate and other isolated burials dating to the first half of the 2nd century are known from intramural sites on this side of the settlement (RCHM 1928, pl 55). It is at this period that London evidently suffered a decline (Marsden 1980, 110–17), and in these circumstances the official boundary may have been ignored in some instances.

In London, the method of marking out boundaries delimiting the *pomerium*[3] is not known. However, in 1977 excavations at Dukes Place, Aldgate, revealed a ditch dug *c* AD 120 which lay near and on a similar alignment to the later wall, over a distance of at least 50 m (see Fig 95; Maloney 1979, 293–4, 296). Since it was too small to have been defensive, was not apparently used for drainage, and was unlikely to have been a quarry, the possibility arises that it marked a boundary in this area (Maloney 1982). It may be significant that a similar ditch, in a comparable position, and on a different stretch of the circuit was found at the nearby Crosswall site (Egan *et al* 1981). When the position of the intervening gate (at Aldgate, see below) and the neighbouring cemeteries (Fig 92) are also considered, it is possible that on this part of the eastern circuit, at least, the wall respected an earlier boundary.

Given its apparent status as the provincial capital by the early years of the 2nd century, it is probable that there were monumental gates or arches astride the main roads into London as they entered the urban area, such as are known from other towns (Hobley, this volume). No extensive investigations of the sites of the gateways have been possible, although details sufficient for a reconstruction of a Roman gate at Newgate have been gleaned from various observations. Here there was evidence for a double carriageway 10·5 m wide, flanked by two substantial rectangular towers (Norman & Reader 1912, 294–5).Curiously, the plinth of the gate is about 1·4 m above that of the adjoining stretches of the landward wall, which led to the suggestion that it was a later addition. Conversely, it has been pointed out that on the basis of style and construction it could equally have been a free-standing monumental arch of earlier date than the wall, as in the case of the Balkerne Gate at Colchester (Marsden 1980, 124). The difference in level could be due to the elevated position of the road on high ground overlooking the valley of the River Fleet to the west.

The course of the wall is markedly irregular and it has been observed how it changes direction in the vicinity of the gateways at Newgate, Bishopsgate, and Aldgate (see Fig 100; Merrifield 1965, 101). If the wall was constructed in lengths round from the east to the west (below, p 100), it follows that from the south-west corner of the fort the course chosen for the wall was the shortest possible route to the Thames via Newgate. At Aldgate, the gateway is in a defensively poor position, being in effect prominently exposed due to the convergence of the wall on either side. From Aldgate to Bishopsgate the wall changes direction twice, perhaps to avoid an obstacle.[4] Although no definite evidence can be offered it is suggested that on these three sites monumental entrances may have already existed before the construction of the wall. An existing structure which certainly influenced the course of the wall was the fort at Cripplegate, since its north and west walls were incorporated into the circuit (Fig 92). Details now available of excavations at the Tower of London in 1956–7 demonstrate that the existence of a substantial stone building on the site of the White Tower resulted in a change of direction in the course of the wall, in order that the building be included within the circuit. Also in this area, it appears that a timber building of Hadrianic date was deliberately dismantled to allow the construction of the wall on its site (Parnell forthcoming).[5]

It was only on the west, south of Newgate, that the natural terrain afforded the defences any distinct benefits, and full advantage was taken. The wall was constructed so close to the ridge overlooking the valley of the River Fleet that a stretch apparently became unstable, tilted outwards, and cracked (Norman & Reader 1912, 288–9; Marsden 1972, 842–4).

Description

Some 33% of the circuit of the landward wall has been recorded or observed (Merrifield 1965, 298–325, map) and a few impressive stretches have been preserved and are accessible (Ordnance Survey 1981). The wall was about 3000 m long and enclosed an area of slightly more than 133 ha; this was a vast circuit and no other town in Roman Britain had defences enclosing such a large area. Indeed in these terms perhaps only six towns in Gaul and Germany exceeded it in size—Nîmes, Vienne, Trier, Autun, Avenches, and Tongres. Calculations of the materials necessary for its construction clearly indicate the magnitude of the undertaking: over 30,000 cubic metres of stone and flint were required, more than 1,000,000 facing stones must have been dressed, and some 500,000 tiles used (Marsden 1980, 127). The tiles were probably made not far from London (*ibid*, 95; Cotton & Laws 1980), but the stone and flints were brought from quarries around Maidstone in Kent, a distance of 70 miles by water. A Roman sailing barge with a cargo of Kentish ragstone (a coarse limestone) from these quarries was discovered on the bed of the Thames near the western end of the wall and may well have been bringing material for its construction (Marsden 1967, 41). The riverside was apparently left undefended at this time although special measures must have been taken to prevent either end of the circuit from being outflanked. Perhaps towers were used to turn extensions of the landward wall along the river-banks for a distance.

The uniform design of the wall and the fact that no reused material seems to have been employed[6] are clear indications that it was planned and built as a single structure. Observations and excavations during the last century have shown that the defensive circuit also consisted of an external V-shaped ditch and, internally, a bank of earth (Fig 93). Evidence of a road bordering the bank has been found

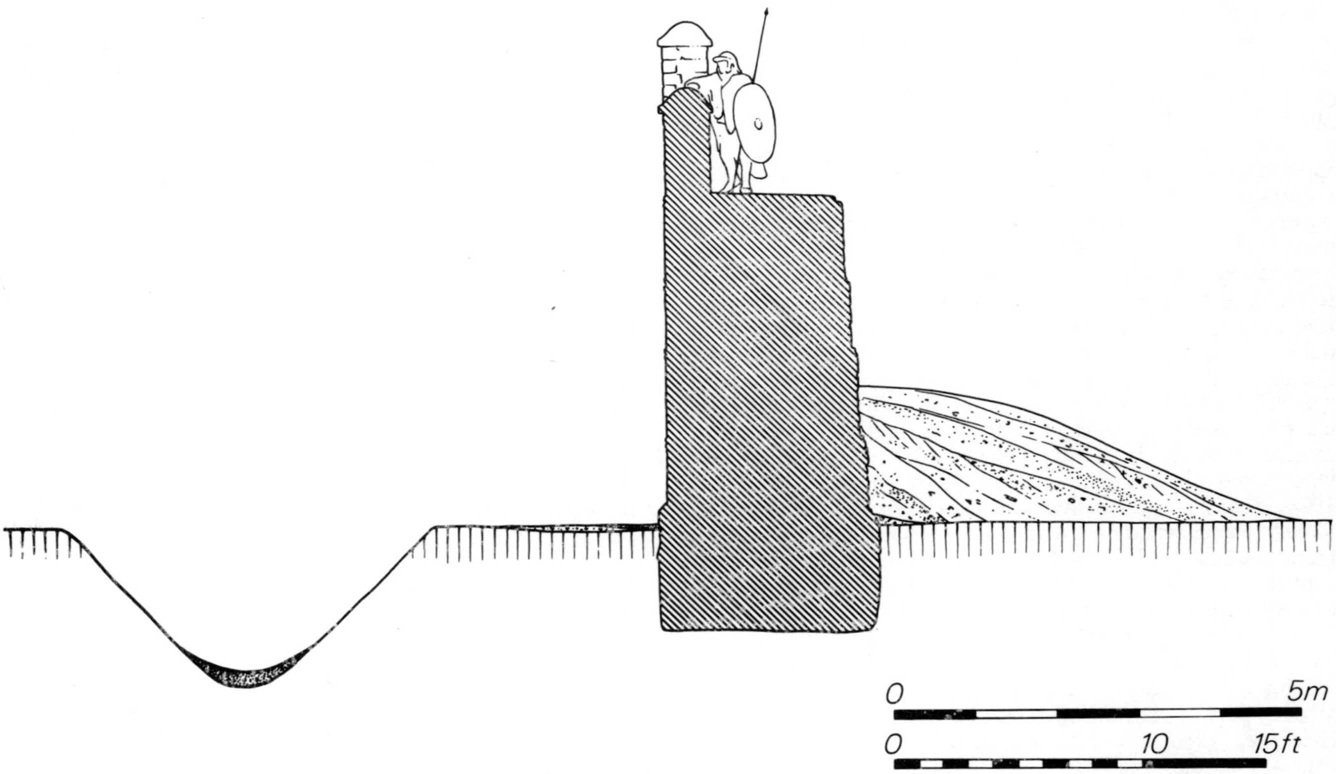

Fig 93 London: reconstructed section through the defences of the landward wall (drawing: C Green ©)

(Grimes 1968, 51; Parnell forthcoming) but if an intramural road did exist it was apparently not continuous (Marsden 1965, 135; 1970, 4, fig 3). Where roads breached the circuit at Aldgate, Bishopsgate, Newgate, and Ludgate, gateways controlled the thoroughfares. Presumably volume of traffic would have dictated that the main gateways be fairly substantial as at Newgate, but of the other gates little is known and those slight Roman remains that have been recorded—for instance at Aldgate (Marsden 1969, 20–3)—are not dated.[7] The remains of a Roman gate at Aldersgate postdate the wall (below, p 110). When the north and west walls of the fort were incorporated into the circuit their respective gates were retained, but since the walls were less than half the width of the landward wall a thickening was added to bring them up to strength (Grimes 1968, 47–9). Four small internal towers (turrets) of one build with the wall have been found (see Fig 100), and these presumably contained timber staircases providing access to the parapet. Their dimensions vary, and there is no apparent evidence of regular spacing, but obviously others existed. The best known example (Marsden 1970, 2–6; for the others, see Merrifield 1965, 299–301, W3, W5, W10) was found between Newgate and Ludgate, measured 2·33 × 2·92 m internally, with walls 0·92 m thick, and contained important evidence for the dating of the wall (below, p 104).

Construction

The construction of the wall conforms to a standard pattern (Fig 94) except for minor details which apparently relate to the nature of the local terrain, or where individual lengths of wall can be identified (RCHM 1928, 72) which were probably the work of different gangs of builders. A study of the variations in construction along different stretches suggests that these lengths were of the order of 25 m (Maloney forthcoming).

The foundation is composed of puddled clay with flints (or occasionally ragstone) laid in a trench which is usually the same width as the base of the wall (about 2·6 m) or slightly wider. Its depth ranges between 0·88 and 1·17 m (RCHM 1928, fig 10) according to the stability of the subsoil: on a site near a stream the foundations were reported as resting on massive piles (Merrifield 1965, 304, W18) and also by Aldersgate (MS notes by A Oswald).[8] At the top of the trench is a footing, usually of two or three courses of concreted ragstone rubble, and at this stage of construction provision was made for culverts to pass through the wall (Merrifield 1965, 302–12, W12, W29, W30, W48/W49). All the known culverts were formed with tiles and two are recorded as having been bonded with red mortar (*opus testaceum* ?). Most had rectangular channels—the largest being 0·76 m high × 0·46 m wide—and apparently drained out of the settlement. However, one (W30) was substantially bigger—1·07 m high × 0·99 m wide—and comprised an arch made of 50 tiles. The drain extending from it was estimated to have had a total length of 55 m and where it terminated outside the settlement five iron bars, fixed perpendicularly, blocked the end. In view of its size and position, this culvert may well have channelled a principal feeder of the Walbrook stream into Roman London.[9]

Above the footing on the external face of the wall, a constant feature is a chamfered plinth of red ferruginous sandstone (in blocks 0·3–1·06 m long × 290–440 mm wide from quarries in the same region as the ragstone and flint), which is usually at or just above the contemporary ground surface (see Fig 101). On the internal face at this point is a triple facing course of tiles, generally with the uppermost tile slightly offset back (Fig 94). Examination of a stretch of wall at Dukes Place revealed a singular aspect of its construction which affected the level of the plinth (Maloney 1979, 293–5). The ground was prepared by the dumping and levelling of a large amount of natural brickearth,

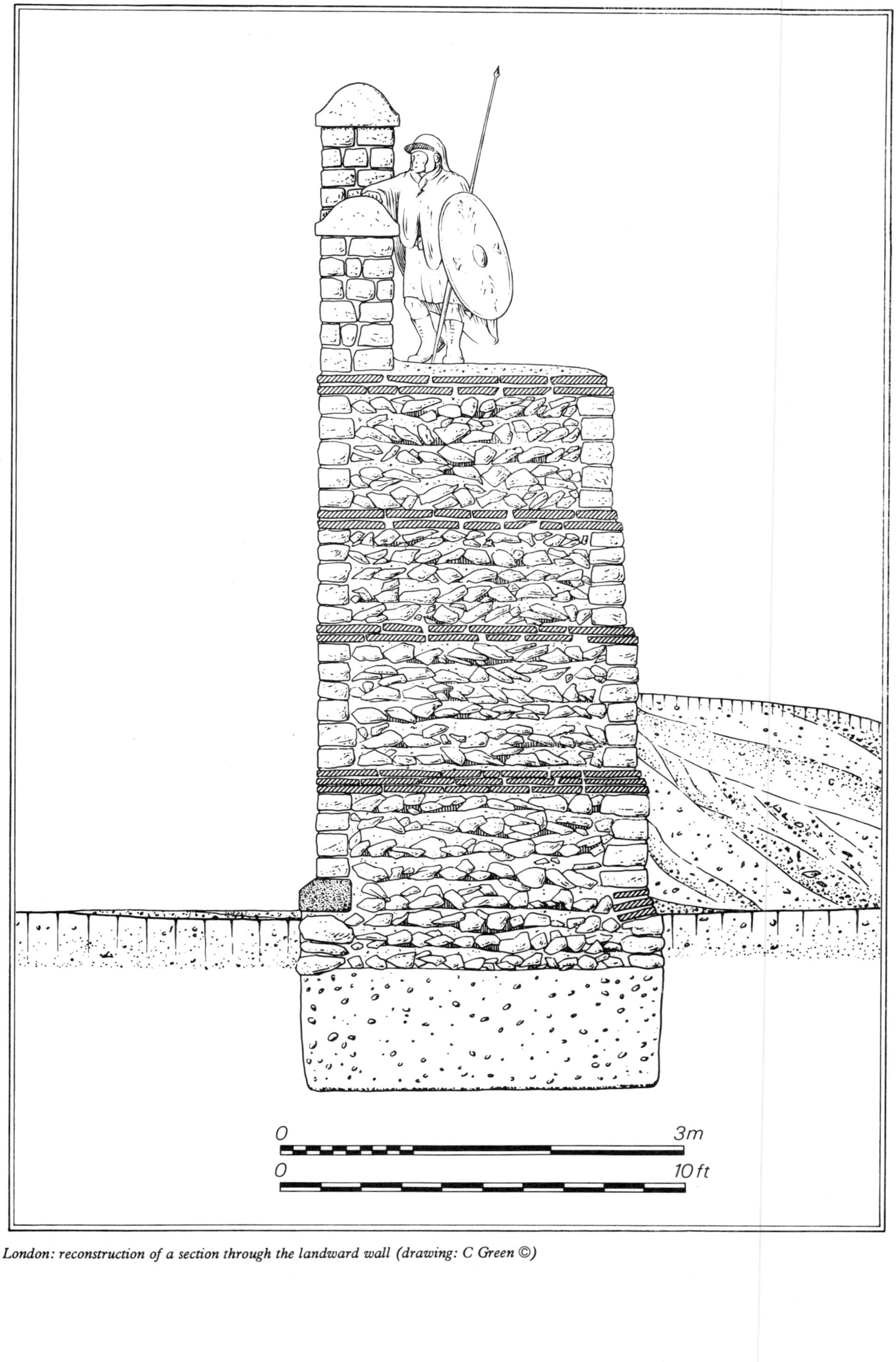

Fig 94 London: reconstruction of a section through the landward wall (drawing: C Green ©)

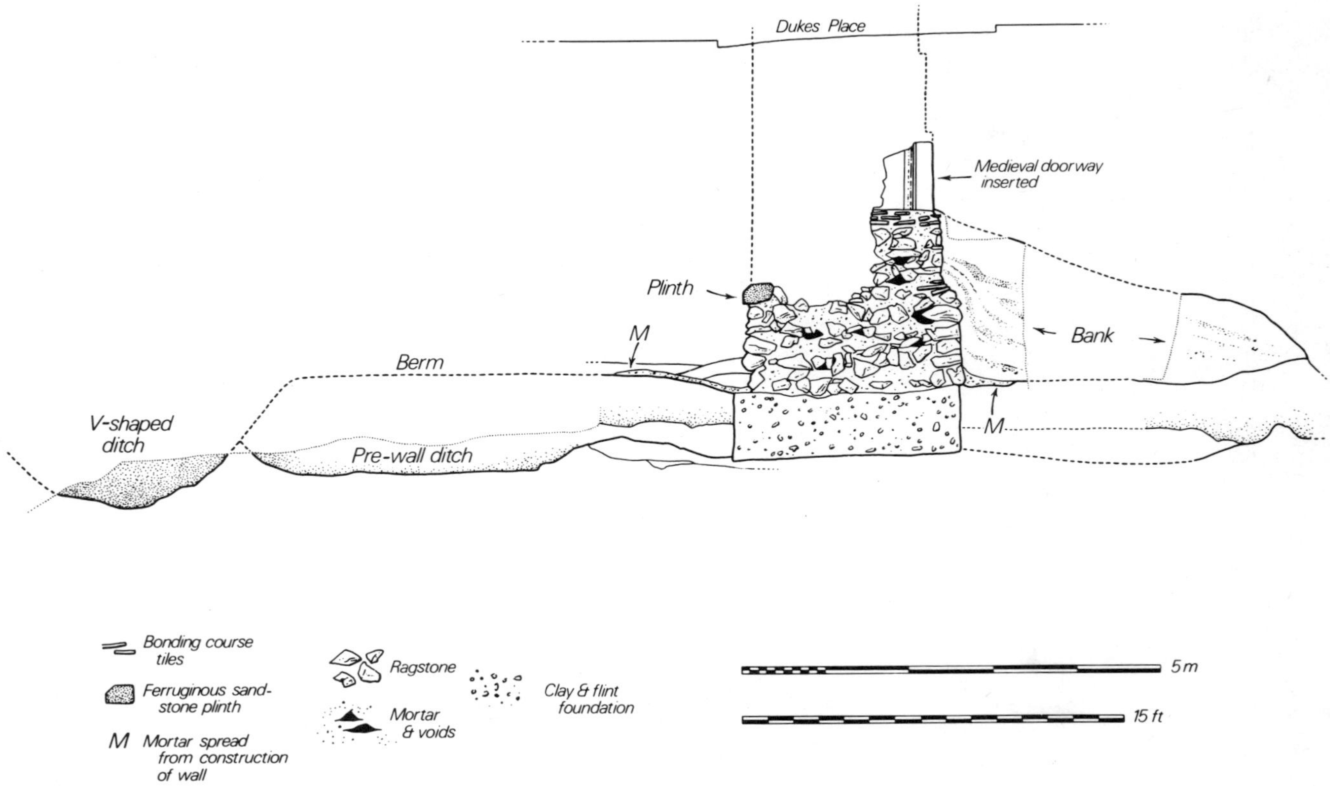

Fig 95 London, Dukes Place: composite section through the defences of the landward wall (drawing: C Green ©)

through which the foundation trench was cut (Fig 95). Set not quite squarely above the clay and flint foundation was a ragstone rubble footing, 1·15 m high, supporting the main body of the wall. Due to this excessively large footing, the plinth lay well above the contemporary ground surface. Similarly, the bottom of the internal bank, instead of being level with the triple facing course of tiles opposite the plinth, lay far below. Presumably the ground surface in this immediate area was so unusually low (a natural hollow?) that even after a substantial dump of brickearth had been deposited, the plinth at its predetermined level along this stretch of wall was inevitably elevated for a distance. Later accumulations containing 3rd and 4th century pottery (above the mortar spread on the external face) make it clear that the level of the ground surface was not raised to that of the plinth in the Roman period. But the wall was not always built up from a set level: at the Tower of London a stretch is known to have been constructed on a slant along the north-south slope down to the River Thames (Parnell forthcoming).

Above the plinth the width of different stretches of wall is known to vary between 2·05 and 2·63 m (RCHM 1928, 72). At this point the width is generally reduced (Fig 94), and thereafter successive reductions occur at the levels of the bonding courses of tiles: normally on the internal face the uppermost tile is offset back 7·6 mm. The main body of the wall comprises well coursed internal and external faces of roughly squared blocks of ragstone retaining a ragstone rubble core, at intervals capped by double or triple bonding courses of tiles. The height of the facing courses varies between 120 and 180 mm and the blocks of ragstone range from 120 mm square to 400 × 180 mm. The joints between are well pointed, usually with a distinctive off-white mortar. In the core the ragstone rubble was laid in courses, sometimes rather haphazardly, between each of which was a layer of extremely hard concrete (*opus caementicium*) apparently poured in since it only partially percolated through the rubble, leaving cavities (RCHM 1928, 71; Maloney 1979, 294). This concrete has been analysed and found to contain a proportion of lime to sand of 1 : 2 (Terry 1900, 354–5). It has been observed that occasionally the courses of ragstone rubble are set in a herringbone pattern (RCHM 1928, 71), but a longitudinal section through the core of the wall at Dukes Place (and close inspection during its demolition) revealed many courses consistently set at an angle which indicated that this stretch was constructed from east to west (Fig 96).[10] In the longitudinal section different mixes of concrete were also apparent (Maloney 1982). Above the level of the plinth the bonding courses of tiles ran through the full width of the wall providing horizontal stability. An example of galletting[11] is perhaps an indication of the importance attached to keeping them quite level (Merrifield 1965, pl 41). Tile courses occur at intervals of 1 m or less depending on the number of intervening facing courses, which varied on different stretches. A number of whole tiles were recovered from Dukes Place and found to be of the standard Roman size, 295·7 × 431·7 mm, that is $1 \times 1\frac{1}{2}$ Roman feet. They ranged in thickness from 31·7 to 50·8 mm, and in colour from orange to dark red, although a yellow tile was recovered, as elsewhere (RCHM 1928, 71). Broken tiles were also used, particularly over the core. At Dukes Place an unusual feature of another stretch of wall, briefly observed, was an offset at a tile course on the external face. A similar example was recorded at Crosswall in the last century (Fig 97), and since a culvert passed through the footing of this stretch (Merrifield 1965, 302, W12) perhaps the external offsets were meant to provide additional stability. Alternatively, it may be yet another sign that the wall was constructed

Fig 96 London, Dukes Place: longitudinal section through a stretch of the landward wall showing details of construction—view looking south (photo: J Bailey)

in lengths by different gangs of builders, because stretches on other parts of the circuit where culverts have been found do not appear to share this feature.

The original height of the wall is not known but stretches have been found existing to the level of the fourth tile course, 4–4·4 m above the plinth (Fig 98; Merrifield 1965, 105). In Canterbury a stretch of the Roman town wall appears to have survived intact to the top of the crenellation, 7 m above the foundation; if so, it is the only Roman crenellation known from Britain (Tatton-Brown 1978, 81). If both defensive walls were of a comparable height, and allowing for a breastwork some 2 m high, the London stretches which survived to the fourth tile course may have been at, or near, parapet level (Fig 94).

At Dukes Place at the junction of the foundation and superstructure there was a layer of mortar on either side of the wall (Fig 95). A similar feature seen on Tower Hill was described as a mortar spill (Merrifield 1965, fig 4), presumably resulting from the construction of the wall. However, a recent excavation on a nearby site found extensive remains of such a mortar layer said to have been repaired with rough stone surfaces where worn thin, which led to the suggestion that it was a hard-standing facilitating the work of the masons (Whipp 1980, 49–50). The absence of putlog holes for scaffolding presumably means that they were carefully blocked up.

A feature of the Roman defences apparently contemporary with the wall was a V-shaped ditch (Fig 93; Merrifield 1965, fig 12). The profile is known to vary somewhat but the ditch is usually 3·05–4·88 m wide, 1·17–2 m deep, and separated from the wall by a berm of 2·7–5·2 m. The possibility of a multiple ditch system has not been investigated because the outer edge of the V-shaped ditch and a considerable area beyond are invariably obliterated by a large ditch of medieval date. The V-shaped ditch has been recorded at various points on the circuit (Merrifield 1965, 105; Maloney 1979, 295) and was recently found at Crosswall (see Fig 101; Maloney 1980b, 68). However, it has yet to be positively located on the western side. There is evidence that the ditch was well maintained and the culverts draining in its direction make it likely that certain stretches at least were 'wet'.

A bank of earth against the internal face has been found on several sites and was quite clearly formed after the construction of the wall (Marsden 1970, 3–4; Maloney 1979, 295), though probably not long after (Whipp 1980, 50; Parnell forthcoming). The bank was not therefore an earlier defensive feature cut back to accommodate the wall. Normally the bank seems to have been up to 2 m high, tailing off some 4 m away from the wall (Fig 95), but on the eastern slope towards the Thames it appears to be 9 m or more in width (Whipp 1980, 50; Parnell forthcoming). Doubtless the material used in its formation was the upcast from the foundation trench and perhaps the ditch; the dumps of predominantly 'clean' natural deposits successively tipped against the wall (Fig 93) may indicate that the bank was formed in stages perhaps corresponding to those in the construction of the wall.

Dating and discussion

London is in the fortunate position of having the construction of the landward wall quite securely and closely dated.

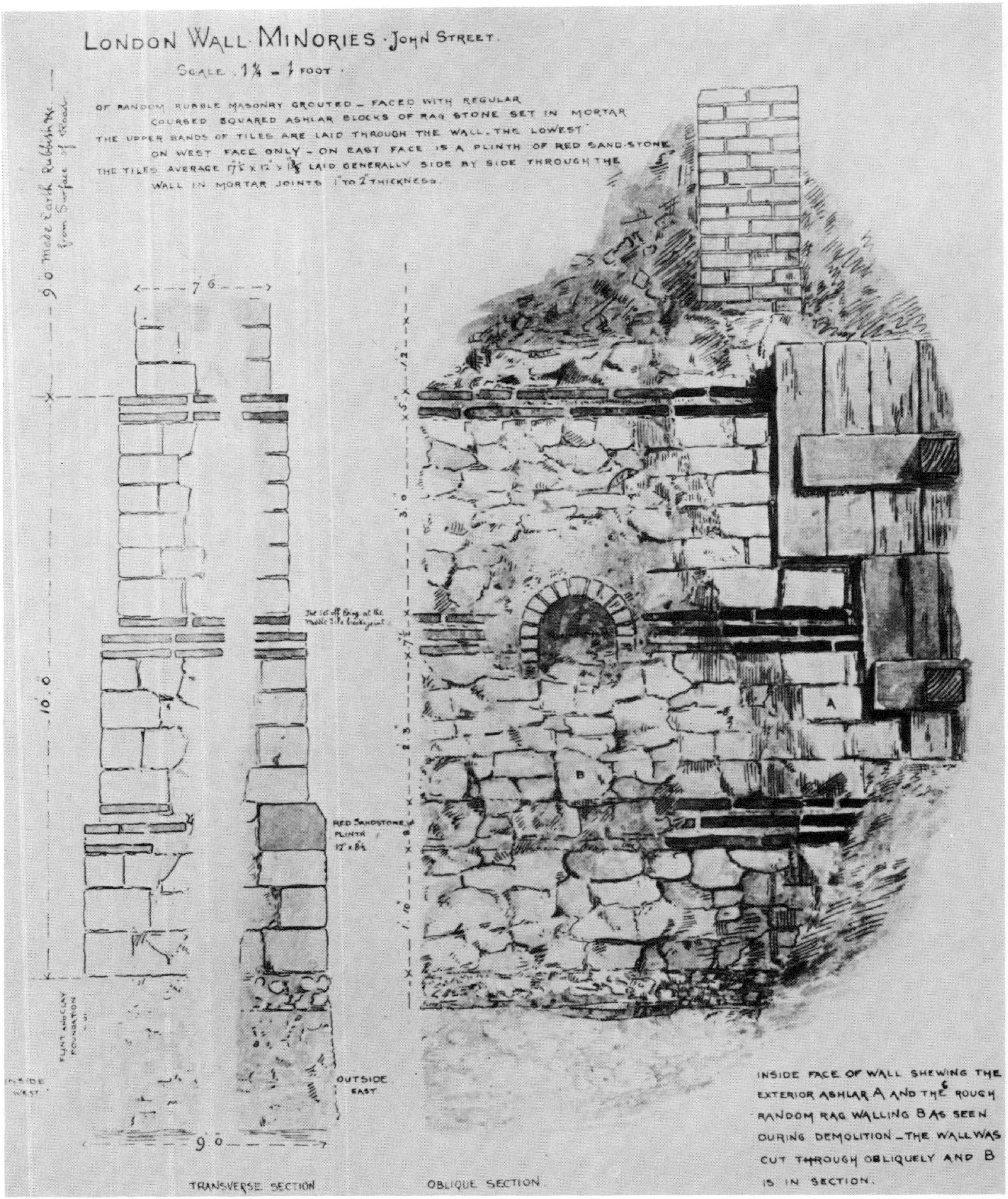

Fig 97 London, Crosswall (formerly John Street): a stretch of the landward wall unusual in having offsets on its external face—recorded on a site just to the south of the recent excavation (from a drawing by H Hodge of 1881, Guildhall Library)

Fig 98 London, Tower Hill: internal face of the landward wall with the original Roman face surviving only in the bottom right-hand corner, between the double tile course at plinth level (modern ground level) and the triple tile course above. The facing work at the higher levels is medieval but to the left of the white bands a further three double tile courses can be discerned in the Roman core of the wall (photo: J Bailey)

Recent excavations have produced pottery from deposits sealed by the wall, and from within the bank, of which the latest sherds date to the last quarter of the 2nd century (Maloney forthcoming; Parnell forthcoming). A coin of Commodus minted in AD 183–4 (but quite worn) was found in a deposit beneath the thickening added to the west wall of the fort (Grimes 1968, 50–1). It can reasonably be suggested therefore that the construction of the thickening was contemporary with that of the landward wall. The crucial *terminus ante quem* was provided by a group of coins in a layer which also contained early 3rd century rubbish, on the floor of an internal tower of one build with the wall (Marsden 1970, 2–6). The coins evidently constituted a forger's hoard since they were found together with *terracotta* moulds made from unused coins of AD 201–10, 210–12, and 215. The moulds had been disposed of with a genuine new *denarius* of AD 213–17 and three other coins issued between AD 210 and 217, suggesting that the forger was at work within a few years of AD 215 and probably not later than AD 225 (Marsden 1980, 126). Thus, assuming that the hoard was dumped in the internal tower about this time, the defences of the landward wall must date to the period AD 175–225. Indeed, if the evidence of the coin of Commodus is accepted, a date after *c* AD 190 must be assumed.

A contentious issue is the reason why these defences were erected. It is a commonly held view that town walls constructed before the mid 3rd century were built for 'decorative purposes, for the sake of civic dignity or, at most, for police purposes. They were certainly not built for military reasons' (Luttwack 1978, 168). There are however specific objections to such a dogmatic assessment in the case of London. To begin with, if status or civic dignity were primary considerations the landward wall is more likely to have been erected in the Flavian–Hadrianic period rather than at the end of the 2nd century. Between times London suffered the effects of a decline from which it was still recovering in the 3rd century (Marsden 1980, 110–19). In the circumstances this would seem an improbable time at which to engage on such a massive undertaking (above, p 97) which would have imposed a severe burden on human, natural, and financial resources. Also, in the latter half of the 2nd century, the construction of defences without the prior permission of the provincial governor and the emperor had been prohibited by law—the two entries in the *Digest* (I, 8, 9 (Ulpian); L, 10, 6 (Modestinus)) were enacted during the principate of Marcus Aurelius, AD 161–80. Thus the construction of the landward wall and associated defences must have been a response to a situation in which they were considered a necessity. Indeed several towns in the south received the protection of earthwork defences in the second half of the 2nd century (Wacher 1966) and London's stone wall may have been part of a wider scheme of defence. It would seem clear that at this particular time the provision of defences had important political overtones for the province, related to contemporary events. Of the known historical events cited, the period of anarchy following the death of Commodus, when Clodius Albinus was governor and a claimant for the Empire, has been seen as the most likely context (Merrifield 1965; 50–1; Frere 1974, 251). The circumstances are well known and there is no need to review the case here, except to point out that it has the merit of allowing adequate time for the large-scale effort needed for the organization of London's landward defences. The wall was clearly carefully constructed but, given the evidence of work by gangs of builders and the resources available to the governor in the provincial capital, the task could have been comfortably completed between the years AD 193 and 197.

Reorganization in the late Roman period

A signal station on the Thames

In 1974 excavations at Shadwell, 1·2 km east of London on the highest point of a natural gravel ridge overlooking the Thames, revealed a free-standing structure 8 m square with walls 2 m wide which was interpreted as a tower (Johnson 1975). Although solidly built the tower may have been erected in haste since, instead of terracing the natural north-south slope chosen for its site, the northern foundations were simply made deeper than those to the south which were buttressed by way of compensation. Apparently this later led to its collapse. The initial phase of construction included a gravel surface and a row of large postholes, which may have been a palisade, 6 m south of the tower. Later, two parallel ditches were dug, the inner ditch replacing the row of posts. The remains of clay floors and sill-beam foundations with uprights, together with

Fig 99 Trajan's Column: a watch tower/signal station and (far left) barrack block on the banks of the River Danube. The wooden platform may have been for use as a beacon with the two stacks providing the combustible material (drawing C Green ©, after C Cichorius)

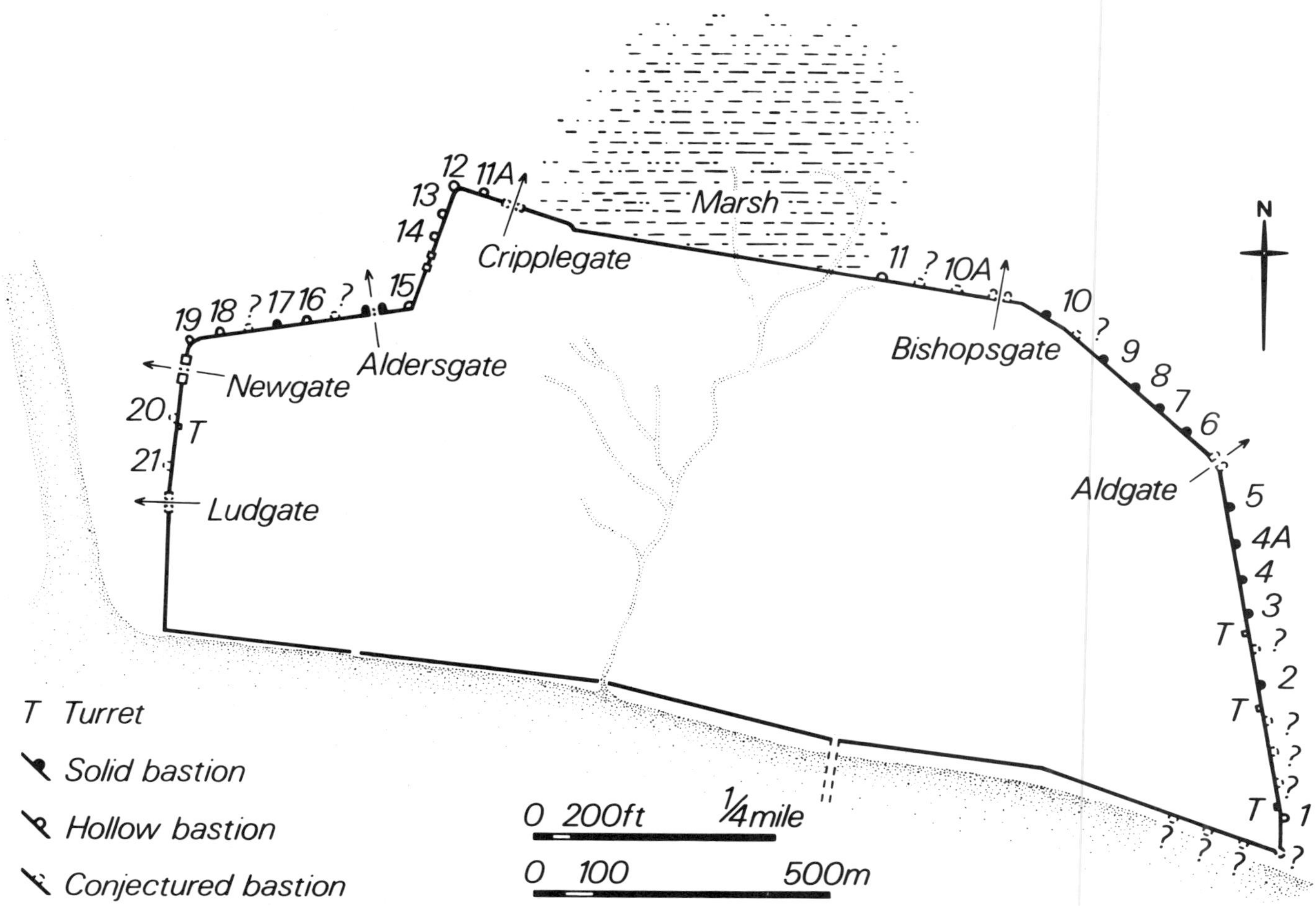

Fig 100 London: plan showing the known and predicted locations of external towers (bastions) (drawing C T Maloney)

burnt wattle and daub from walls, indicated the presence of timber buildings which were apparently contemporary with the tower. Other features included a huge hole dug on the site of a spring, which fed an elaborate water system consisting of oak-plank drains packed with clay and a water storage tank similarly constructed. A leather bikini and a coin of Allectus (AD 293–6) were found in the tank and a distinctive bronze strap pendant from a military belt was also recovered. The site yielded more than 300 coins, the majority of which were of the period AD 260–95 with those of Carausius and Allectus predominating. The water system was renewed *c* AD 330–40 and the site was probably abandoned soon after AD 360.[12]

The ground plan, massive walls, palisade, and later ditches suggest that the excavations at Shadwell revealed the site of a signal station or watch tower and the timber buildings and military equipment may indicate that it was garrisoned, perhaps on a temporary basis (Fig 99). The dating evidence points to construction towards the end of the 3rd century, at a time when the Saxon Shore forts were being organized. The tower at Shadwell would have been clearly visible against the skyline to observers on the eastern side of the landward wall, and may have been at the head of an inland signalling system connecting London with the coastal defensive system (Johnson 1975, 280). Obviously London, being on a navigable river and without defences on the riverside, was particularly vulnerable to attack from Saxon raiders.

Bastions (external towers)[13]

The bastions in London are all evidently later additions to the landward wall, but form two groups both geographically and in methods of construction (Fig 100). The eastern group contain Roman monumental stones reused as building material and have solid bases (see Fig 104).[14] The western group are hollow (except B17) and are not known to have incorporated reused stone from Roman monuments. Curiously, between the two groups is a 230 m stretch of wall on the north along which no bastions are known. It is certain that at least B11A of the western group is of medieval date and B14 and B15 are of similar construction (Grimes 1968, 71–6). Also, early medieval worked stones are reported as having been found in B16 (Fox 1891, 610). Therefore, the western group are generally regarded as being medieval and the eastern group Roman, although the facts of the matter may have been more complicated.

Recent excavations have provided more evidence about the eastern series of bastions. At Crosswall the site of a previously unknown bastion (B4A) was discovered (Figs 101, 102; Maloney 1980b, 68–76). Much of the foundation survived together with a group of stones from the superstructure which were the remains of an inscribed Roman tombstone, probably of early 3rd century date (Fig 103). A close examination of the face of the wall above showed that B4A had not been bonded to it (Fig 101). In plan, the foundation was roughly rectangular with rounded corners on its outer face. It was composed of layers of ragstone,

Fig 101 London, Crosswall: the external face of the landward wall, V-shaped ditch, and foundation of Bastion 4A (photo: T Hurst)

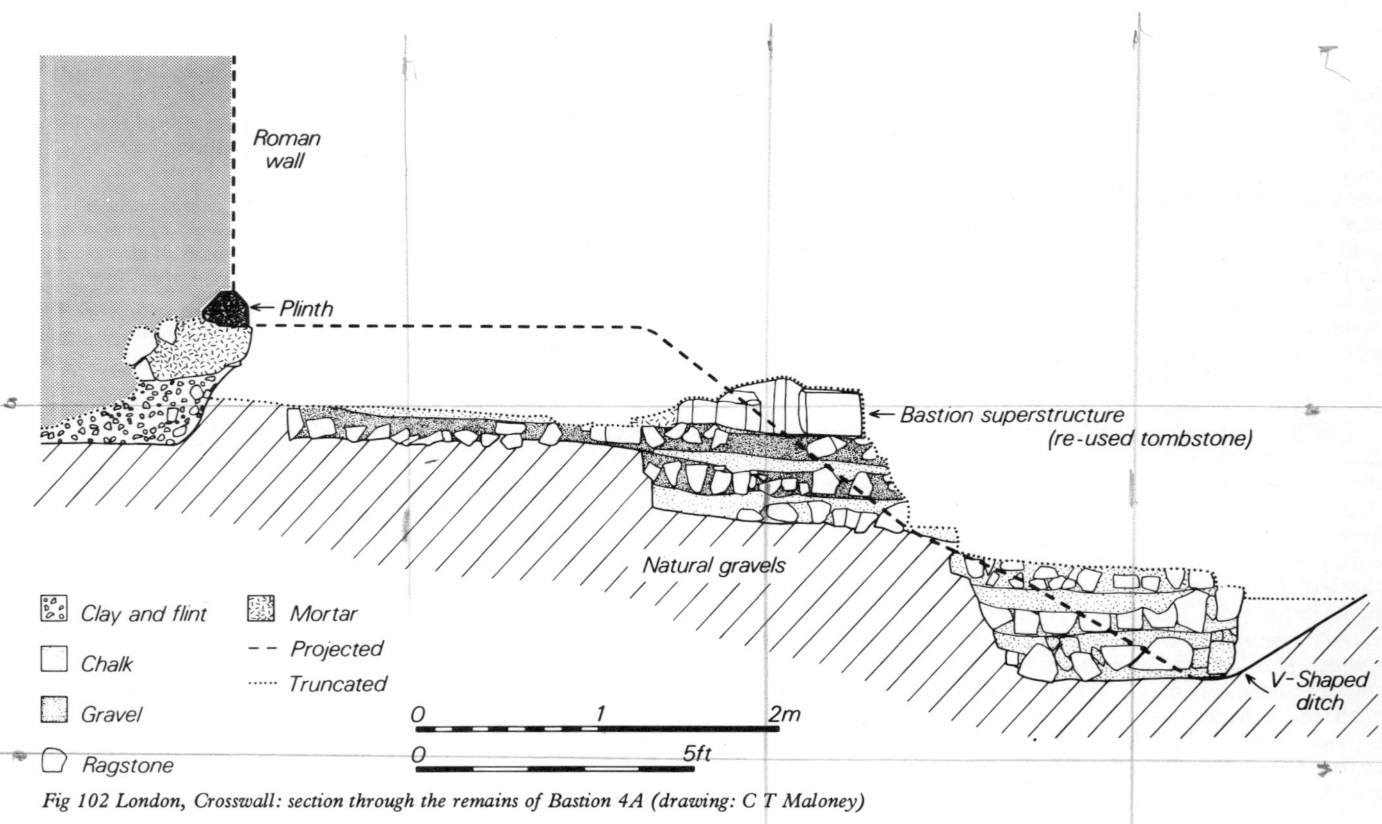

Fig 102 London, Crosswall: section through the remains of Bastion 4A (drawing: C T Maloney)

Fig 103 London, Crosswall: fragments of an inscribed tombstone recovered from Bastion 4A (photo: J Bailey)

Fig 104 London, Bastion 6: showing its solid construction and a coping stone in situ *(photo: P Marsden)*

flints, and lumps of *opus signinum* embedded in gravel and capped with rammed chalk. On the upper levelled surface on which the tombstone lay there remained the imprints of other large blocks of stone indicating that the superstructure was of the D-shape usual amongst the eastern group (Egan *et al* 1981). Although it is clear that the V-shaped ditch had been backfilled prior to the construction of the bastion, its foundation was carefully stepped down in two stages into the bottom of the ditch (Fig 102) implying that the builders were aware of the existence of that feature. The demolition of the bastion not later than the 13th century, together with the other evidence, suggests that it was Roman. Its discovery has important implications for the rest of the eastern group because, whereas previously they appeared to be irregularly spaced (Merrifield 1965, 111), the discovery of B4A indicates otherwise. It becomes clear that the bastions were spaced at intervals of nearly 53 m, and where longer gaps occur in the series the sites of other bastions yet to be found can be predicted (Fig 100; Maloney 1980b, 70–2).

This element of regular spacing implies that the eastern group were contemporary and belonged to a single system, so that dating evidence from any one should indicate the date of the whole group. At Dukes Place the V-shaped ditch was also found to have been deliberately backfilled, presumably to allow the construction of nearby B6, and the fill material contained 1st and 2nd century pottery, a barbarous radiate coin of the late 3rd century, and a coin of Constans of AD 341–6 (Maloney 1979, 295). The area around Bastion 6 had been previously excavated and a deposit which had accumulated against its outer face contained pottery and bronze coins of the period AD 364–75 (Marsden 1980, 172). Therefore it can be proposed that the eastern group may have been constructed between AD 341 and 375.

The eastern group share several common characteristics as well as their solid bases and the reuse of monumental Roman stone. They are generally D-shaped (Fig 104), 5·8–7·9 m wide, and project 4·4–5·6 m from the wall (Merrifield 1965, 320–5). The use of ragstone, pink mortar,[15] chalk, and flint are recurring features; in several instances (B6, B7, and B9) courses of bonding or facing tiles, characteristic of Roman work, are recorded, and this may have been a common aspect. Although they were apparently carefully erected, there is striking variation in the details of construction to the extent that each may have been the work of an individual gang of builders.

The most obvious feature of the construction of the eastern bastions is the reuse of Roman funerary monuments, and architectural and coping stones; this is a practice well attested in late Roman fortifications elsewhere (Blagg, this volume). These large blocks may have been

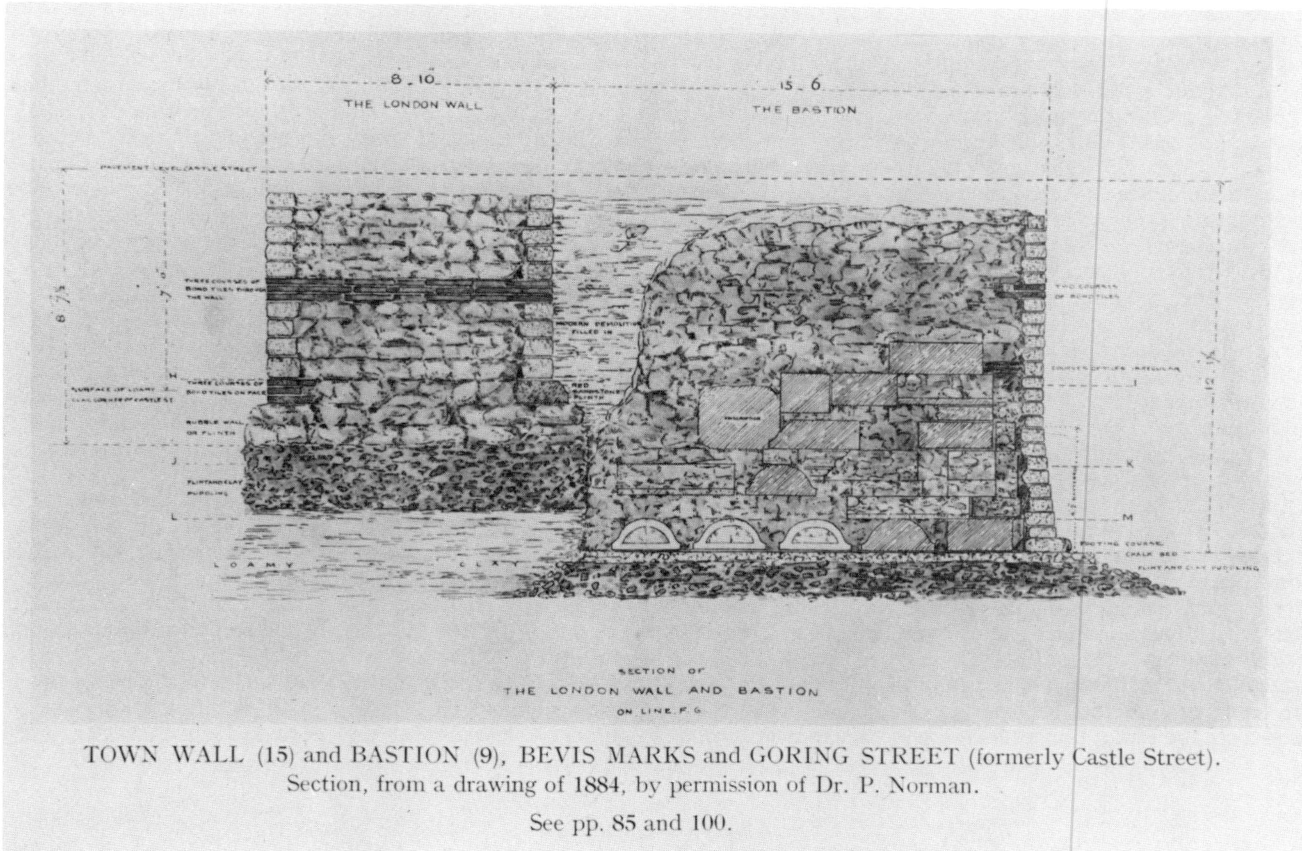

Fig 105 London, Bastion 9: section showing details of its construction and relationship to the landward wall (from a drawing by H Hodge of 1884)

Fig 106 London, Bastion 9: coping stone with raised platform and rebate, and raised platform with sockets (50 cm scale; photo: J Edwards)

Fig 107 London, Bastion 9: L-shaped coping stone with raised platform and rebate (50 cm scale; photo: J Edwards)

deliberately used to provide a substantial footing, because where detailed records have been made (B9, B10, and B11) it is apparent that they were employed up to a level just above that of the plinth in the adjoining stretch of wall (Fig 105; G Marsh, pers comm). In two instances (B9 and B10) the bastions had solid rubble cores above, up to their surviving heights of 3·29 m and 2·92 m respectively. Since there are no indications that the bastions were situated opposite internal towers (Fig 100) it is not clear how access was gained unless it was directly from the parapet walk.

Of the tombstones recovered from the bases of the bastions the best known is that of the famous historical figure, the procurator Iulius Classicianus (Cottrill 1936). The rest of the worked stones are about to be studied in detail, but an important discovery recently made is that parts of the same monument have been identified in both B8 and B10 (G Marsh, pers comm). Since B8 and B10 are not neighbouring bastions, an implication of this discovery is that the blocks were stockpiled before use. A preliminary look at the coping stones has led to some interesting general conclusions (Figs 106, 107; R Merrifield, H Chapman, G Marsh, pers comm). Given the number of reused tombstones presumably from the adjacent cemeteries, the coping stones could have come from the enclosing walls or even individual monuments. However, since they are consistently 0·60 m wide and up to 1·75 m in length it seems strange that cemetery walls sufficient to support them have not yet been detected. It is perhaps more likely that they came from the top of the crenellation of the landward wall, from those parts which had to be dismantled when the bastions were constructed (Fig 108). However, some of the coping stones are L-shaped (Fig 107) and even T-shaped, and must have come from structures more complicated than the parapet wall. Also, some of the bastions are known to have contained many more coping stones than could have come from a related part of the crenellation. It has therefore been suggested that when the bastions were built the landward wall was modified in other respects, and certainly this seems likely with regard to the gates. If designed on the Newgate model, those on the eastern side at least would presumably have needed to be rebuilt in the late Roman style with projections in line with those of the bastions. It has generally been assumed that the Roman bastions served as platforms for light artillery (*ballistae*), and perhaps the solid construction of the eastern series is indicative of this function (but see Baatz, this volume). Nevertheless, an important implication of their regular spacing was to provide enfilading fire along the face of the wall (Fig 108) and in a unified system gate towers must have played their part. At Aldersgate, on the west, a gateway is known to have been inserted in the landward wall and apparently featured projecting towers such as are found at Saxon Shore forts and other late Roman fortifications (Fig 109). The use of tile bonding courses and pink mortar are further pointers to its Roman date (Merrifield 1965, 102). Intriguingly, the nearby west gate of the fort which continued in use with the landward wall was later blocked, possibly at this period (Grimes 1959), and its

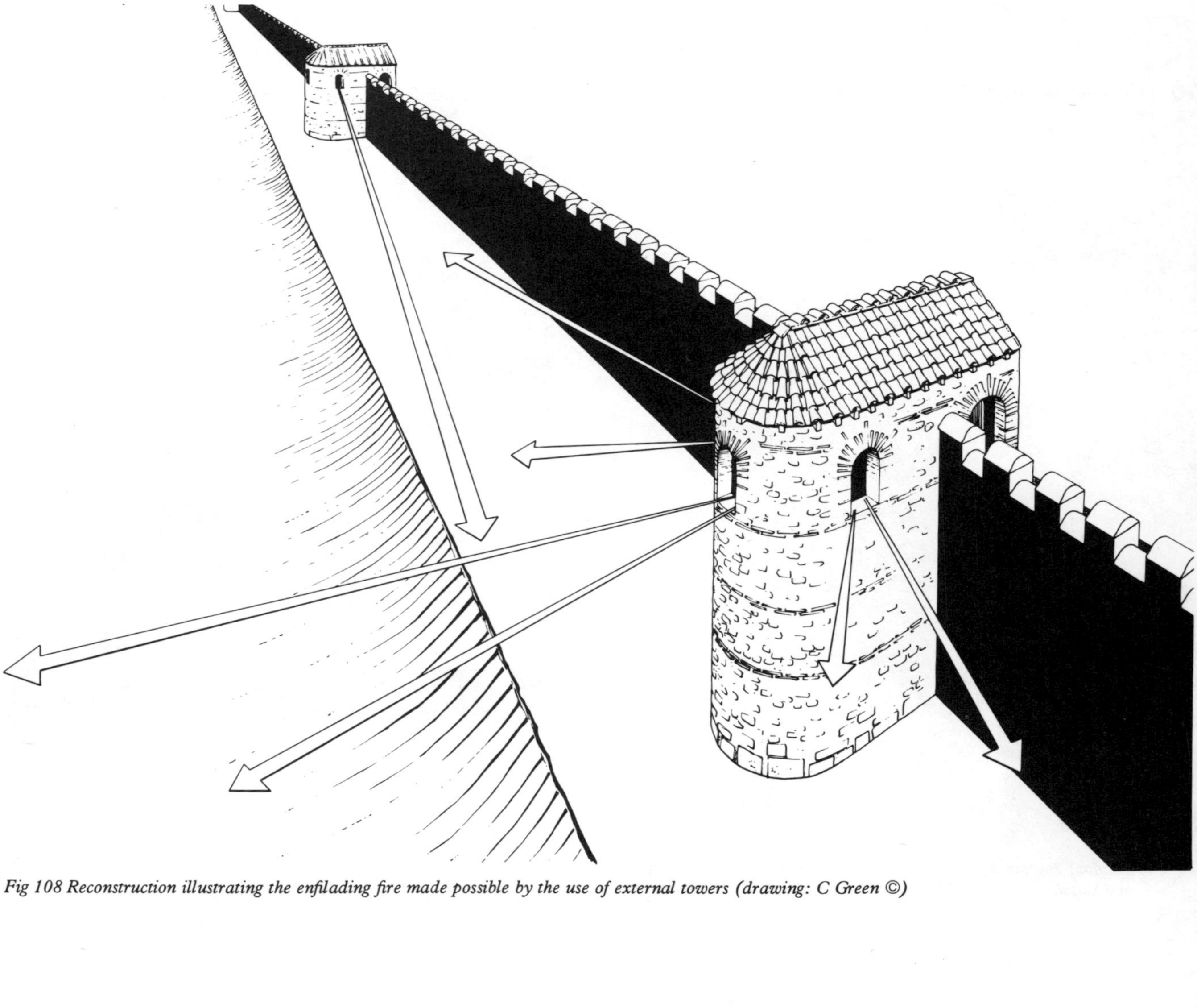

Fig 108 Reconstruction illustrating the enfilading fire made possible by the use of external towers (drawing: C Green ©)

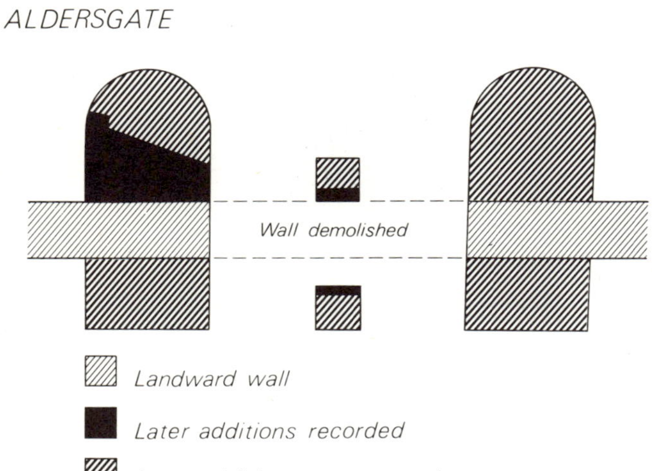

Fig 109 London, Aldersgate: ground-plan of the remains of a projecting gate tower and two piers which were a later addition to the landward wall. The west gate of the 4th century 'Saxon Shore' fort at Pevensey is shown for comparison (drawing: J Maloney)

disuse may be related to the construction of the boldly projecting gate towers at Aldersgate.

The backfilling of the V-shaped ditch and construction of the bastions necessitated the digging of a new, wider ditch in a more advanced position (Fig 110), as found by Corder (1959) at Great Casterton. The inner edge of such a 4th century ditch was recorded at Dukes Place where the correspondence of the base of the bastion with the slope of the berm of the ditch indicated that both were in use contemporaneously (Marsden 1980, 172). Once again the medieval ditch had virtually obliterated its predecessor but excavations at Ludgate in 1974 proved more informative (Hobley & Schofield 1977, 44–5, fig 10). There a wide flat-bottomed ditch of late Roman type containing pottery provisionally dated to this period survived to more than 1 m deep and 6 m wide, and its actual width could have been as much as 13 m.[16] A ditch of similar character recorded at St Alphage (Grimes 1968, 86, fig 21), Houndsditch, and Dukes Place (Maloney & Harding 1979, 350–1, 353) may be part of the same system; in the latter two examples single sherds of pottery in the upper fills were of early medieval date but a large late Roman ditch could have remained open for a long period.

The riverside wall

Before 1975, one of the most controversial aspects of Roman London was the question of the existence of a defensive riverside wall. A 12th century reference seemed unequivocal: 'there runs continuously a great wall and high . . . with towers along the North at intervals. On the South, London was once walled and towered in like fashion, but the Thames . . . has in the course of time washed away these bulwarks . . .' (Butler 1932, quoting William Fitzstephen). However, despite the documentary evidence no conclusive archaeological proof was found until excavations at Baynards Castle near the south-west corner of Roman London revealed the remains of a riverside wall over a distance of nearly 115 m (Hill *et al* 1980, 27–76). On the eastern part of the site a 38 m long stretch was examined (Fig 111); this was built on a foundation of five neat rows of oak piles 0·20 m square by 2–2·6 m long capped by a thick chalk raft. The inner face comprised four courses of ragstone set in a rough herringbone pattern surmounted by a tile course above which the wall was offset back 100 mm. Then followed five courses of ragstone and this time a double tile course with a much wider offset of over 300 mm; this pattern was duplicated above. Flat tiles were used as in the landward wall but also roofing tiles (*tegulae*) set with the flanges downwards into the mortar. The tile courses apparently did not run the full width of the wall which was gauged to be 3·0 m, the outer face having been eroded away by the Thames, as Fitzstephen had asserted. The core was of rubble concrete formed from ragstone, chalk flint, tile fragments, and lumps of *opus signinum.* Between the second and third tile courses the herringbone work of the internal face gave way to courses of quite well dressed rectangular blocks, more reminiscent of the landward wall. This work was completely covered up by a clay bank against the internal face which survived to a height of 2 m. Mortar spreads in the bank at the level of the tile courses may indicate that it was used as a working platform, being successively added to as the wall gained in height. A culvert 200 mm square and capped by a flat tile was built on a diagonal through the wall and a corresponding channel was created through the clay banked behind. No evidence of an external ditch was found and the proximity of the River Thames presumably made it unnecessary.

Further west there was a change in the construction of the wall; the foundation was no longer piled but instead consisted of large blocks of ragstone rammed into the subsoil. This difference appears to have coincided with a change in the natural terrain, gravels giving way to clay in this area. Once again the tiles were only used as facing courses, this time without offsets. The wall on this part of the site seems to have been deliberately demolished and the absence of a bank may be related to this event. However, the most interesting difference in construction was the reuse of 52 monumental blocks of stone (Figs 112, 113),[17] the majority of which were identified as being from a monumental arch and screen of deities (Hill *et al* 1980, 124–93). They were apparently used in an upper part of the internal face of the wall at which point there was a wide offset of 0·7 m. Evidence that these blocks of stone probably came from redundant buildings on adjacent sites has come from a recent excavation nearby (Williams 1982, 26–8). It has been suggested that the wall may have been of more than one build (Marsden 1980, 174, 177). Clearly

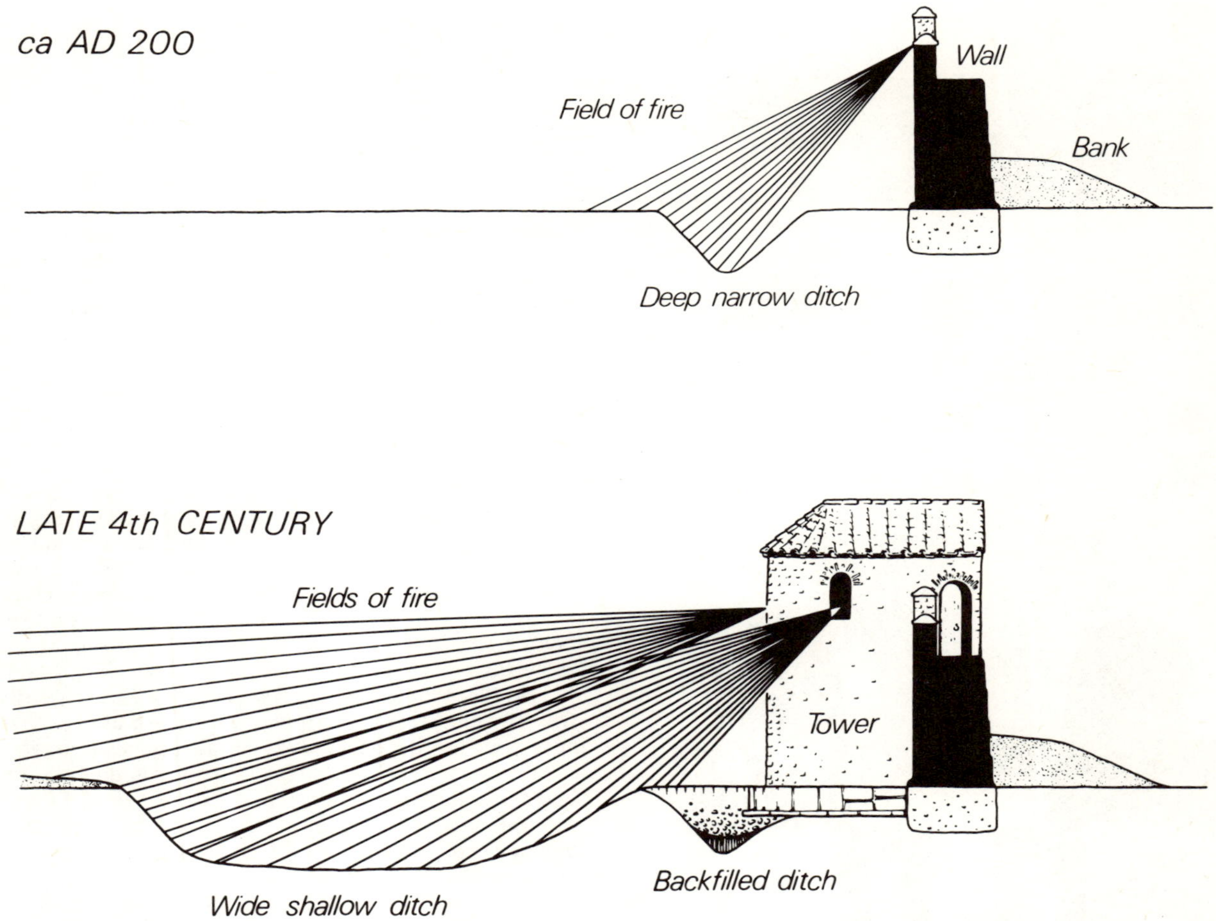

Fig 110 Reconstruction illustrating the effects of the addition of external towers to a defensive circuit (drawing: C Green, after P Corder)

the construction of the riverside wall was not as regulated as that of the landward wall, but the differences observed need not imply any more than individual gangs of builders set to work on different stretches, as suggested for the construction of the bastions. Indeed, there are many similarities in details of construction between the riverside wall and the bastions, which point to their being associated. The dating evidence for the riverside wall lends support to this connection: a date between AD 350 and 370 for the foundation piles has been indicated by radiocarbon analysis although recalibration would push this later into the 4th century (Hill *et al* 1980, 88–93).

The existence of the riverside wall as a continuous feature from the south-west to the south-east corner of Roman London was subsequently confirmed by excavations at New Fresh Wharf (Miller & Schofield forthcoming), and at the Tower (Parnell 1978). In both instances the method of construction was comparable to that recorded on the eastern stretch of wall at Baynards Castle, although minor differences in construction were apparent at the Tower.

A later modification to the south-east corner of the riverside wall

Excavations at the Tower also revealed a riverside wall of later date than that founded on oak piles, and quite different from the other stretches (Fig 114; Parnell 1977). The ground having been levelled with a layer of gravel, an insubstantial footing was formed of ragstone, chalk, and flint set in puddled clay. The internal face of the wall was well coursed and consisted mostly of dressed blocks of ragstone, supplemented with Purbeck marble, tufa, sandstone, chalk, and tile. The external face was only accessible in a limited area, but clearly did not feature a plinth; in addition to well coursed blocks of ragstone, a double tile course was noted at 1·1 m above the base. The wall was 3·2 m wide and the core was composed of courses of ragstone rubble, together with chalk, tile, and *opus signinum*, which alternated with layers of gravelly mortar. Timber lacing was detected in the core, parallel and near to the internal face; it appeared to be associated with an offset 1·50 m above the base, at which level a number of putlog holes were visible. There was no sign of corresponding lacing near the external face although a lateral bracing timber was observed. Where the wall changed direction, reused large stone blocks were employed to turn the corner through a 100° angle. The pointing on the internal face of the wall was not weathered owing to the presence of deposits dumped up against it, evidently very soon after its construction. Material from this bank included a comprehensive selection of late Roman pottery and over 30 coins, the latest in date being of Valentinian II, of the period AD 389–92.

The alignment of this wall is significant: it was constructed 4 m to the north of its predecessor but did not extend to the landward wall (compare Figs 92 and 100). Instead, some 14·5 m to the west it turned abruptly south to meet the earlier riverside wall which is assumed to have continued to the corner. To the west the later wall is presumed to have been directly joined to its forerunner at some point. However, the earlier wall remained upstand-

Fig 111 London, Baynards Castle: the riverside wall with its foundation of oak piles and chalk raft—eastern side of the site. Over half the width of the wall was eroded away by the River Thames during the Saxon period (photo: T Hurst)

Fig 112 London, Baynards Castle: a stretch of the riverside wall which collapsed towards the River Thames, with the monumental stones exposed in the foreground—western side of the site (photo: T Hurst)

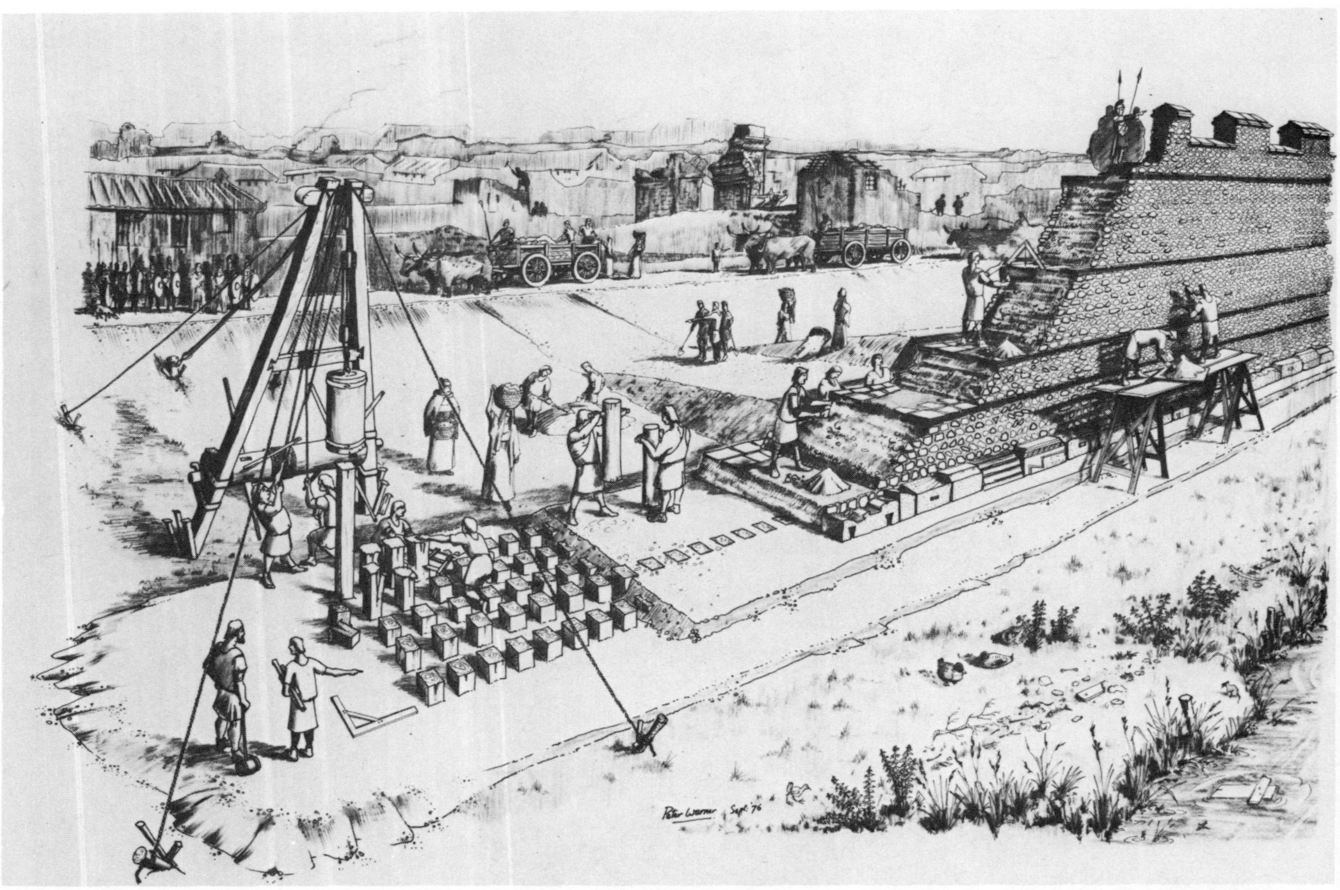

Fig 113 London: reconstruction of the building of the riverside wall based on evidence from the Baynards Castle site (drawing: P Warner)

Fig 114 London, Tower of London: the late 4th century addition to the riverside wall with the Tower curtain wall above. Bottom left, the riverside wall changed course to the south, towards the River Thames, and is assumed to have adjoined part of the earlier riverside wall on the far side of the curtain wall (photo: D Craig)

ing to the west of the point where it was met by the later wall and therefore overlapped it for a distance. The effect was to make the south-east corner more prominent, suggesting a promontory linked to the landward defences with concealed access through a narrow channel on the western side.

Discussion

The construction of the defences *c* AD 200 was evidently directed against a landward attack, but during the late Roman period attention became focused on the threat of seaborne invaders coming up the River Thames. The 3rd century signal station at Shadwell appears to represent the first step in a reorientation which continued in the 4th century with the addition of bastions and the construction of a riverside wall to complete the circuit (Fig 115). Furthermore, the modification to the riverside wall at the very end of the Roman period suggests that the south-east corner was especially strengthened to form a stronghold. This then is the proposed outline of events which will now be discussed.

The construction of bastions on the eastern side of the landward wall raises the question of their apparent absence on the remainder of this circuit. Clearly, the eastern side was most vulnerable to attack from downstream and the natural terrain on this side in particular afforded no advantages to defenders. However, to the north the existence of an extensive marsh outside the wall probably made the provision of bastions unnecessary. The west side was less accessible to such an attack; doubtless the bridgehead would always have been well defended and, once the riverside wall had been erected, the western side would have been even more sheltered. Yet an attempt seems to have been made to reorganize the defences on the western side as shown by the existence of a late Roman ditch at Ludgate. Therefore, although the eastern group of bastions appears to be a unified series, nevertheless they may have been part of an unfinished scheme. Perhaps bastions in Britain were only provided where considered absolutely necessary—London apparently has more than any other town (Butler, this volume). Certainly the riverside wall is an impressive indication of the resources that could be called upon as necessary, and its construction during this period must indicate that a comprehensive reorganization of London's defences was undertaken. There is some evidence to suggest that this also involved the remodelling of certain of the gates to serve the needs of the new defensive system and perhaps the blocking of Cripplegate and its replacement by a gate of the appropriate design in a better position at Aldersgate. Further, it has been argued elsewhere that the riverside wall itself may have had its own series of bastions (Maloney 1980b, 73–4). A possible context for such a reorganization is the well documented Theodosian reconstruction which followed the *barbarica conspiratio* of AD 367. It may be no coincidence that the signal station at Shadwell was abandoned about this time. Even if not contemporary, as suggested by similarities of construction, the erection of the bastions and riverside wall during the late period must have been part of a coherent strategy.

It is clear that the defence of London was considered

Fig 115 London: reconstruction of the defensive circuit c *AD 370 (drawing: A Sorrell)*

of paramount importance, for a remodelling of the south-eastern corner was carried out with the erection of an extremely well constructed and substantial wall after AD 390—the latest known date for the structure of this kind in Britain. This extra strengthening implies the establishing of a stronghold on that spot and circumstantial evidence has been offered in support of this proposition (Maloney 1980a, 60). Perhaps in this late development a link can be seen with the creation of citadels in some of the towns of Gaul (Butler, this volume). The date of this phase of refortification may have coincided with the last known official expedition to Britain under Stilicho at the very end of the 4th century, and is perhaps a reminder of the importance of London to the administration of the Empire even at that late period.

Acknowledgements

As will be apparent from the 'personal communications' and the bibliography, this paper owes much to the work of many people and not least all those who helped with the excavations. I am particularly indebted to Geoff Marsh, and to Hugh Chapman, Ralph Merrifield, and Geoffrey Parnell for the benefit of discussions with them. Comments made by Tony Dyson and Harvey Sheldon greatly improved the text, and Geoffrey Parnell and Dave Bentley kindly supplied information in advance of their publications. Chris Green and Cath Maloney produced the splendid drawings and Trevor Hurst, Jon Bailey, and Jan Scrivenor the excellent photographs. Finally, I am most grateful to Cath Maloney for her assistance.

Notes

1 '...for the rebels, avoiding fortifications and places under military protection, and eager for booty easily won, sought only what was most worth the plundering and was unguarded by defenders'.

2 Owing to the lack of a detailed study of Roman burials found in London, the remarks made in the text must be regarded as tentative. In Fig 92 isolated burials have not been plotted, only areas which have groups of three or more and presumably were part of an officially recognized cemetery. The extent of the areas so marked is not related to density of burials but simply an indication of the localities where groups have been found.

3 The *pomerium* defined the official boundary of a settlement and originally was closely associated with an ancient religious ritual observed on the foundation of a town. A furrow was ceremonially ploughed to mark the limits and a ditch dug along its line, with the earth piled on the inner edge to form a bank. The open space left on either side of the ditch and bank was referred to as the *pomerium* (Reid 1913, 29–30). However, in Britain the *pomerium* 'might be no more than a slight bank or even a line marked only by boundary stones, and such an arrangement could be considered adequate even when the main approaches were equipped with monumental arches or gates' (Rivet 1964, 78).

4 A site just east of Bishopsgate—inside and adjacent to the wall—presents two possibilities (Merrifield 1965, 286, site 317): firstly, there is the group of (Antonine ?) cremation urns previously referred to (p 97) and, secondly, at a higher level, a mosaic pavement was found reportedly 1·07 m from the wall. At the Tower of London, a Roman building which was skirted by the wall was left similarly close to it (see p 97).

5 There may have been another example of such an occurrence at Aldgate (Chapman 1973, 10).

6 A stretch of wall examined in 1982 at Ludgate is perhaps an exception because a lump of pink mortar and broken tiles were discovered in its foundation.

7 It has been strongly suggested that there was a Roman gate on the eastern part of the circuit, south of Aldgate (Merrifield 1965, 316, G1), but no definite evidence has been found.

8 Over part of the site the wall apparently crossed a marsh and in this area the foundation was widened to 4·42 m and reinforced with massive blocks of ragstone, as well as being supported on piles.

9 From the existence of culverts on the northern part of the circuit (W29, W30, and W33), it has been concluded that three main streams of the Walbrook were channelled into London (Merrifield 1965, 89). However, one of these culverts apparently drained out of the settlement (W29) and another is presumed from the presence of a large drain which is not known for certain to have passed through the wall (W33). The suggestion that the culvert (W30) which seems most likely to have carried a feeder of the Walbrook was replaced at a higher level later in the Roman period may be misleading since there must be a suspicion that the same feature was recorded during separate sightings.

10 The core of the wall recently revealed at Ludgate (see note 6) appears to have been constructed from north to south, which would accord with the theory that the wall was built from the east round to the west. However, a more detailed examination is needed before this can be confirmed.

11 Galletting is the insertion of small stones or flint flakes into mortar while it is still wet. For instance, where a course of stonework was uneven, gallets would be used as miniature wedges to reduce the thickness of what would otherwise have been a very wide mortar joint.

12 I am grateful to Mike Hammerson for information about the coin evidence.

13 In London, towers added to the external face of the wall have invariably been referred to as bastions, eg Wheeler in RHCM 1928 and

Merrifield 1965. I have followed this convention for the purposes of cross-referencing although the use of 'bastion', and indeed 'turret', may be misleading (see Butler, this volume). However, while not strictly accurate, these terms are perhaps less confusing than references to projecting, external, or interval towers, as well as internal and gate towers.

14 It has been assumed that B1 and B11 were originally hollow (Merrifield 1965, 68), but since they were used and possibly rebuilt in the post-Roman period in circumstances in which it would have been preferable for them to be made hollow, this assumption should be regarded with caution.

15 The pink mortar apparently used for securing one of the large blocks in the footing of B4A to its foundation has been visually identified as *opus testaceum*.

16 Excavations during 1982 on the same site (see notes 6 and 9) confirmed that the primary fill of the wide, flat-bottomed ditch contained late Roman pottery and that the ditch was apparently recut in the Saxon period (P Rowsome, pers comm). It has been suggested that a late Roman addition to the bank behind the landward wall on a nearby site may have been the upcast from the digging of this wide ditch (Marsden 1980, 173).

17 Fig 113 is a reconstruction of work in progress on the piled stretch of the riverside wall, but the reuse of blocks of masonry forming a plinth on the external face is a supposition.

References

Bentley, D, forthcoming The Roman cemetery at St Bartholomew's Hospital, *Trans London Middlesex Archaeol Soc*

Butler, H E (trans), 1932 William Fitzstephen's *Description*, in *Norman London* (F M Stenton), 26–32

Chapman, H, 1973 Excavations at Aldgate, 1972, *Trans London Middlesex Archaeol Soc*, **24**, 1–56

Corder, P, 1955 The reorganisation of the defences of Romano-British towns in the fourth century, *Archaeol J*, **112**, 20–42

Cotton, J, & Laws, A, 1980 Excavation round-up 1979, *London Archaeol*, **3**, 14, 386

Cottrill, F, 1936 A bastion of the town wall of London, and the sepulchral monument of the Procurator Julius Classicianus, *Antiq J*, **1**, 1–7

Egan, G, Maloney, C, & Maloney, J, 1981 *Excavations at Crosswall, London EC3, 1979–80*, unpubl (archive report available at the Department of Urban Archaeology, Museum of London)

Fox, G E, 1891 Notes on a recent discovery of part of the Roman Wall of London, *Archaeologia*, **52**, 609–16

Frere, S S, 1974 *Britannia: a history of Roman Britain*, 2 edn

Grimes, W F, 1959 Roman Britain in 1958, I, sites explored: London, *J Roman Stud*, **59**, 126

——, 1968 *The excavation of Roman and medieval London*

Hill, C, Millett, M, & Blagg, T, 1980 *The Roman riverside wall and monumental arch in London*, London Middlesex Archaeol Soc Spec Pap, **3**

Hobley, B, & Schofield, J, 1977 Excavations in the City of London: first interim report, *Antiq J*, **56**, 31–66

Johnson, T, 1975 A Roman signal tower at Shadwell, E1, *Trans London Middlesex Archaeol Soc*, **26**, 278–80

Luttwack, E, 1978 *The grand strategy of the Roman Empire*

Maloney, J, 1979 Excavations at Dukes Place: the Roman defences, *London Archaeol*, **3**, 11, 292–7

——, 1980a The Roman defences of London, *Curr Archaeol*, **73**, 55–60

——, 1980b The discovery of Bastion 4A in the City of London and its implications, *Trans London Middlesex Archaeol Soc*, **31**, 68–76

——, 1982 *Excavations at Dukes Place, London EC3, 1977–8*, unpubl (archive report available at the Department of Urban Archaeology, Museum of London)

——, forthcoming *The Roman landward wall and defences of London: excavations 1977–1982 and recent research*, London Middlesex Archaeol Soc Spec Pap

Maloney, J, & Harding, C, 1979 Dukes Place and Houndsditch: the medieval defences, *London Archaeol*, **3**, 13, 347–54

Marsden, P, 1965 Archaeological finds in the City of London, 1962, *Trans London Middlesex Archaeol Soc*, **21**, 2, 135–9

——, 1967 *A Roman ship from Blackfriars, London*, Guildhall Mus monogr

——, 1969 Archaeological finds in the City of London, 1966–8, *Trans London Middlesex Archaeol Soc*, **22**, 2, 1–26

——, 1970 Archaeological finds in the City of London, 1966–9, *ibid*, **22**, 3, 1–9

——, 1972 Mapping the birth of Londinium, *Geogr Mag*, September, 840–5

——, 1976 Two Roman public baths in London, *Trans London Middlesex Archaeol Soc*, **27**, 1–70

——, 1980 *Roman London*

Merrifield, R, 1965 *The Roman city of London*

Merrifield, R, & Sheldon, H, 1974 Roman London Bridge: a view from both banks, *London Archaeol*, **2**, 8, 183–91

Miller, L, & Schofield, J, forthcoming *Excavations at New Fresh Wharf 1974–8: Part 1, Roman*, London Middlesex Archaeol Soc Spec Pap

Milne, G, 1982 Further evidence for Roman London Bridge?, *Britannia*, **13**, 39–44

Norman, P, & Reader, F, 1912 Further discoveries relating to Roman London, 1906–12, *Archaeologia*, **63**, 257–344

Ordnance Survey, 1981 *Londinium: a descriptive map and guide to Roman London*

Parnell, G, 1977 Excavations at the Tower of London, 1976–7, *London Archaeol*, **3**, 4, 97–9

——, 1978 An earlier Roman riverside wall at the Tower of London, *ibid*, **3**, 7, 171–6

——, forthcoming The excavation of the Roman city wall at the Tower of London and Tower Hill, 1954–76, *Trans London Middlesex Archaeol Soc*, **33**

Reid, J S, 1913 *The municipalities of the Roman Empire*

Rivet, A L F, 1964 *Town and country in Roman Britain*

RCHM, 1928 *An inventory of the historical monuments in London*, **3**, *Roman London*, Royal Commission for Historical Monuments

Tatton-Brown, T, 1978 Canterbury, *Curr Archaeol*, **62**, 78–82

Terry, J, 1900 On a fragment of the Roman wall discovered in the Old Bailey, *Trans London Middlesex Archaeol Soc*, n ser, **1**, 351–5

Wacher, J S, 1966 Earthwork defences of the second century, in *The civitas capitals of Roman Britain* (ed J S Wacher), 60–9

Whipp, D, 1980 Excavations at Tower Hill, 1978, *Trans London Middlesex Archaeol Soc*, **31**, 47–67

Williams, T, 1982 St Peter's Hill, *Popular Archaeol*, **4**, 1, 24–30

The function and organization of late Roman civil defences in Britain

Graham Webster

I have rashly undertaken to attempt to find answers to the difficult questions posed by the late civil defences of Britain: what was their purpose and how were they operated?

It would seem self-evident that their main purpose was for defence against threats which made their massive size necessary, with projecting towers at gateways, corners, and at intervals along the walls. The defences in Britain were crude in comparison with the sophisticated systems developed in the eastern provinces. The extra thickening given to walls offered protection against mining and the battering ram, and it also provided space for the operation of siege weapons and for the stacking of missiles consisting mainly, as Baatz has indicated, of stones of difference sizes. Towers and interval chambers provided accommodation for troops either on garrison duty or in transit.

Towards the end of the 3rd century there was a change in the character of the barbarian groups pressing on the north-west frontiers; they had become so large that they were virtually impossible to destroy or turn back. The pressures forcing these peoples to move to the west and south made them all the more determined to find land where they could rest and settle. But there is no evidence that any of these groups were equipped to attack walled settlements. Sieges took time and required technical and engineering skills which the western barbarians did not possess. Nor could they afford to wait too long in any one place when there was always the threat of the sudden appearance of a Roman mobile field army. Possibly the builders of town defences believed that the very presence of such formidable barriers would persuade the invaders to move elsewhere.

But the great wall which Aurelian built round Rome (270–274) had provision for the employment of defensive tactics. A gallery was built on the inner side of the wall to allow defenders to move rapidly to any threatened point. The towers were provided with wide apertures clearly designed for the use of artillery. But, as Todd has pointed out (1978, 82), there is no evidence that the walls of Rome or of any other city were permanently manned by specialist units of the army. Nor is the building of this new type of civic defence in the west associated with any known increase in personnel trained for such artillery. In fact, Marsden has argued (1969, 195) that there was a serious shortage of *ballistarii* in the 3rd and 4th centuries. He draws attention to the caustic remarks of Ammianus (xix, 5,2) about the two Gallic legions at the siege of Amida on the upper Tigris in AD 359. This frontier city had been fortified with walls and towers by Constantine II and provided with artillery (*muralia tormenta*). The legions had been raised by Magnentius in his bid for the purple and after the battle of Mursa were considered suspect, and so despatched to the eastern frontier. It was perhaps unreasonable to castigate legions so recently enrolled for their lack of knowledge and experience in such highly-skilled and technical matters, and they certainly behaved with great courage when put to the test. The army was well equipped with siege weapons, as Ammianus clearly states (xxiii, 4) in his well known description of the army of Julian in his eastern campaign of 363. But these pieces of artillery, collectively known as 'mural engines' (*instrumenta muralia*) (xxiii, 4, 14), were designed only for sieges, in attack and defence. By the mid 4th century most major expeditions were equipped with siege-trains, and in this case Ammianus records that the great Euphrates was hardly wide enough for the thousands of vessels carrying supplies and siege-engines (xxiii, 3, 9). River transport has obvious advantages in the movement of such large quantities of bulky material and objects.

Ammianus may have discussed, in one of the lost books, the smaller types of artillery used in the field when speed and mobility were essential. Vegetius, in his description of troops drawn up in six battle lines (ii, 15) places two types of small arrow-shooters (*manuballistae* and *arcuballistae*) in the front rank. If one accepts Schenk's argument that the *antiqua legio* of Vegetius was derived from the lost work of Tarruntenius Paternus,[1] these weapons may have been in use by the late 2nd century, but it seems more likely that several earlier sources were used. The *arcuballista* was a non-torsion weapon powered by a composite bow of horn, wood, and sinew and had presumably developed from the much earlier Greek *gastraphetes.*[2] Baatz has identified the field-frames found at Orşova and Gornea as belonging to *manuballistae;*[3] these were, however, torsion machines, presumably with a windlass for winding and tightening. The introduction of a new type of all-metal frame allowed the development of powerful weapons which could be carried and operated by one man, hence their appearance in the first line of battle.

There is a passage in Ammianus which may indicate another use of these machines. It concerns Julian's activities in Gaul in 356 against the Alamanni, when he planned a rapid advance to the north from Autun to take them by surprise. To avoid delay he took *cataphractarii soli et ballistarii*, ie 'only heavily-armoured cavalry and artillerymen', which Ammianus had to admit would hardly seem to have been a suitable force (xvi, 2, 5). Whether he was critical of its composition or of its size is not, however, clear.[4] The passage suggests that the artillery unit was able to keep pace with the cavalry, and therefore was not encumbered with animal-drawn equipment.[5] Although the two types of artillery weapons were the lightest pieces available, they would have needed two men to carry them, one for the weapon itself, and the other for its base and missiles. Their great advantage was that they were light enough to be carried by infantry; and it would surely not have been too difficult to find ways of using horses to carry them, either slung on a harness or on a wooden frame attached to the back. The *ballistarii* with their weapons would have been mounted if they had been part of a rapid mobile column.

There is little doubt about the growing importance of artillery in the late Empire, but the weapons required specialist units trained to operate them. These are listed in the *Notitia* as *ballistarii* both *seniores (Or*, vii, 8) and *iuniores* (*Or*, vii, 21; Marsden 1969, 196). There was a legion of *ballistarii*, a *legio pseudocomitatensis* in Gaul (*Or*, vii, 97), and a *praefectus militum ballistariorum* at Bodobrica under the command of the *Dux Moguntiacensis* (*Occ*, xli, 23). The machines were made and presumably serviced at the *fabricae ballistariae*, two of which were in Gaul.[6] But artillery was by no means confined to units so named; two building inscriptions from High Rochester (Bremenium)[7] refer to the building of *ballistaria*, or artillery platforms. They were identifed by Richmond, who also demonstrated their potential field of fire (1940, 98). He pointed out that these massive stone and clay platforms must have been built for heavy machines, although it is very doubtful if they were used for hurling the great stone balls, about a hundredweight each, a few of which have been found near the site. The unit occupying the fort at this period was *Coh I Fida Vardullorum*, a milliary infantry unit, which must have been specially trained to use these machines, with technical officers (*ballistarii*)[8] for maintenance.

The next and more difficult problem to consider is that of the regular army units stationed in towns. There is little doubt that this was indeed a regular practice on the eastern frontier. Amida was garrisoned by *Legio V Parthica* with native irregulars (Ammianus, xviii, 9, 2). Corbulo in the reign of Nero had trouble with the troops who had served their time in towns, and had become physically soft and lacking in discipline (Tacitus, *Ann*, 13.35). It is not surprising to find in the pages of Ammianus many references to troops in the towns, not only in these areas[9] but also in Gaul. Julian found troops garrisoned at Augustodunum (Autun) where the large circuit of walls was at that time in a state of ruin (xvi, 2, 1), but it was the veterans who were most active in defence. Troops were later distributed among the towns as they could be better provisioned there than in forts. While troops were living in the cities they would naturally have been expected to defend them when attacked. It would have been sensible if an armoury of defence equipment was available, since not all the units possessed such specialist armaments.

A possible conclusion can only be reached by inference, rather than from evidence. Town defences were only built when the need arose and this appears to have been purely a military one; the protection of civilians was entirely secondary. A case has been made for the late 2nd century town defences of Britain having been a military scheme to protect a communication network (Webster 1975). The needs in the later period were much the same, strongly defended positions providing bases from which the army could operate. They were placed where provisions could be gathered to enable units to winter or stay for long periods. One should therefore regard the fortified towns as part of the military defence system protecting the province from threats from known directions. The logic of this is that defence equipment would be part of the scheme and this seems to be implied by Ammianus, though it would have seemed so obvious to him that it is not surprising that he fails to record it specifically.

The regular army would have found it impossible to defend all the towns on a permanent basis. Could then the citizens organize a kind of militia? By the late 4th century a complete division had been created between military and civil functions (MacMullen 1963), so it would seem unthinkable that civilians would have been able to form bands of volunteers to man their defences. In any case, only trained army personnel could have maintained and operated the siege equipment. Apart from regular army units there were only two sources from which men could have been drawn for such a purpose who would have satisfied the requirements of the rigid bureaucracy of the late 4th century. Firstly, there were the veterans, such as those who defended Autun (see above). The establishment of *coloniae* had ceased by Hadrian's day in favour of the more flexible arrangement of encouraging veterans to settle in and develop frontier lands, the officers becoming a kind of military squirearchy (MacMullen 1963, 107–8, n34). With their considerable privileges and tax exemptions, they became natural leaders in many communities. Furthermore, their sons by law became soldiers. Thus, veteran families would have constituted the main citizen element in most frontier towns, and elsewhere they are often found grouped together in small communities in a specific part of a town (MacMullen 1963, 109). In time of danger they would have been the first to organize the defences and would have been an obvious group to be entrusted with the operation of the machines, if no regular units were at hand. This is a logical assumption, but it has no supporting evidence from any epigraphic or historical source.

The other source of acceptable manpower would have been the irregular barbarian units enrolled in the army in the form of *laeti* and *gentiles*. Theodosius, in his scheme to provide improved defences for the British cities, may well have installed such units in them. This idea has received some support from Frere who has suggested (1974, 399) that there were troops in certain towns under the command of the *vicarius* of Britain, since his insignia in the *Notitia* consist of walled enclosures symbolizing the five provinces, whereas the *vicarii* of the other dioceses have only the required number of female titular deities. This interesting proposal has been seriously challenged by Tomlin (1976, 195) who considers that prefects and their deputies, the *vicarii*, could only have assumed command in an emergency; it is unfortunate that the lists of *laeti* and *gentiles* in chapter xlii of the *Notitia* fail to include any in Britain. Frere has linked his suggestion of the presence of these irregular units with the discovery of a military type of equipment which has been considered as belonging to Germanic mercenaries (Hawkes & Dunning 1964). However, as more and more examples of these baldric fittings have been found, the original thesis has been subjected to considerable modification (Hawkes 1974) and criticism (Dickinson 1978; Mann 1977, 14). Much of this is due to the great variety of types, the varying quality of craftsmanship, and the fact that pieces have been found in Saxon female graves. There seems little doubt that the decorated belt was a symbol of official status, and in some instances could have been military, but it seems now very doubtful that one can accept the imaginative idea of Stevens (1971) that they mark the distribution of garrisons of *comitatenses*, or even of *laeti* and *gentiles*, without further finds and more intensive study. We are left, therefore, with some attractive ideas, but little firm evidence to support them.

If one turns to the structures themselves, it has been assumed since Corder's paper in 1955 that the towers added to the walls of British towns were for the use of artillery. Unfortunately, none of these survives to a sufficient height to show whether they had openings like those in the towers of Gaul and Rome. Nor is there evidence from Saxon Shore forts, except for the one in a tower on the north side of Pevensey (Johnson 1976, fig 37). In some cases, as at Burgh Castle, the bastions are solid throughout their surviving height of 16 ft, but all have a large hole 2 ft 2 ins in diameter set into their upper surface. It has been suggested that this was for the pivot or turntable of a siege weapon, but Johnson has pointed out (1976, 38, 39, 115) that it would have been quite impossible to direct fire from them along the fronts of the walls. It would also have been difficult to depress the arrow-firing *ballistae* to a sufficient angle to achieve this (Richmond 1930, 79–80, n14; adapted by Baker 1971). The functions of machines placed in or on the towers would have been to provide a frontal field of fire to break up bands of attackers as they approached the counterscarp of the ditch. Those who managed to get to the base of the walls would have been within the range of archers or non-torsion *ballistae*. A fully equipped fort or town would have had several kinds of fixed engines capable of being turned for directing both stones and arrows, as well as the lighter, hand-held weapons. The stone throwing machines needed space and had to be mounted on the tops of towers. Only the lighter arrow-shooters could have been fired from the openings from within the towers.

While one would expect the military sites to be properly equipped, does this also apply to the towns? If the choice of the towns to receive towers was a military one, it would be logical to assume that they would have been provided with siege equipment, but they could only have been operated by army units or veterans. There is no evidence to indicate the kind of organization which may have been created by Constans or Theodosius to deal with this problem. When, however, the Britons were left to their own resources, the position would have been entirely different, and each city would then have had to make the best arrangements possible, regardless of bureaucratic regulations. Army groups had been taken from Britain on several occasions for the defence of the other western dioceses, but few returned. Nevertheless, units must have remained here to continue to defend at least the frontiers of the diocese. There seems no reason to suppose that the northern frontier was abandoned, but that the arrangements made by Theodosius and Stilicho, whatever they were, provided an adequate barrier.[10] The suggestion (Mann 1979) that the large civil settlements in the northern frontier zone declined and that there was a deserted landscape, with the surviving population gathered in the few walled towns like Carlisle, does not accord with the absence of any large-scale land invasion in the early 5th century. Mann explains this (1979, 150) as a waning of Pictish power, yet it was the Picts who were such a menace along the east coast, a situation which was ended only by the Saxon fleet under the mythical commanders Hengist and Horsa. The frontier held because the peoples living there were now defending their lands, and a secondary line of defence had developed from the early 2nd century concept of military families owning the frontier lands. The logical outcome of such an arrangement was the foundation of native kingdoms (Richmond 1958, 125; Frere 1974, 392–3) under romanized rulers, in the style of the old client kingdoms. When the system failed, as it did in North Wales, a suitable ruler was moved into the area with his kinsmen and followers.[11]

While those in control of independent Britain might be able to depend on the security of their frontiers, the town defences now had to be manned by any available army personnel, veterans, or mercenary groups that could be found. In such circumstances strange archaeological anomalies, like the timber building in the Cirencester amphitheatre (Wacher 1964, 18; 1975, 314) and the mausoleum at Kingsholm, Gloucester (Brown 1975) must be expected. Small armed groups and their successors would eventually have tended to take control of the town and its citizens, causing a rapid decline in romanized ways of life, but at the same time ensuring, in many cases, a continuity, however tenuous, into Anglo-Saxon England.

Notes

1 Schenk 1930; Paternus was a praetorian prefect under Marcus Aurelius and one of his field commanders.

2 Marsden 1969, 195, n1; though elsewhere (1971, 6) he states that it was introduced in the 4th century. There is no record of non-torsion weapons for many centuries, but they must have continued in use in the east.

3 Baatz 1978; they are identified with Heron's *cheiroballistra.* Baatz (pers comm) considers that Marsden has misunderstood the figures in the difficult text and reconstructed too large a machine (Marsden 1971, 206–33).

4 He does, however, state that Julian was unable to pursue the fleeing barbarians because he was impaired by his heavy armour (*gravitate praepeditus armorum*).

5 As shown on Trajan's Column, operated by legionaries (lxvi, 162; Richmond 1935, fig 7).

6 *Not Dig Occ*, xix, 33; one of them was at Augustodunum.

7 *RIB*, 1280 and 1281; for *auxilia* using artillery, see *Bonner Jahrb*, **166** (1966), 194 ff.

8 They are listed in the mid 2nd century list of *immunes* (Digest, i, 6, 7).

9 eg Pamphylia, Pisidia, and Lycia had troops quartered in many towns (*per municipia plurima*, xiv, 25). There were troops in the towns of Greece (xiv, 11, 13). Zosimus (ii, 34) makes a general statement that the field armies of Constantine were quartered in frontier towns which had no need for them and that this was detrimental to frontier discipline.

10 The coin series continues down to 410 (Kent 1951).

11 Cunedda's transfer is, however, seen by some scholars as a 9th century myth (Chadwick 1958, 32–4; Dumville 1977, 181–3).

References

Baatz, D, 1978 Recent finds of ancient artillery, *Britannia*, **9**, 1–18

Baker, A, 1971 Viroconium: a study of the defences from aerial reconnaissance, *Trans Shropshire Archaeol Soc*, **58**, 3, 197–211

Brown, D, 1975 A fifth-century burial at Kingsholm, in Excavations at Gloucester: third interim report: Kingsholm 1966–75 (H Hurst), *Antiq J*, **55**, 290–4

Chadwick, N, 1958 *Studies in the early British church*

Corder, P, 1955 The reorganisation of the defences of Romano-British towns in the fourth century, *Archaeol J*, **112**, 20–42

Dickinson, T, 1978 British antiquity: post-Roman and pagan Anglo-Saxon, *ibid*, **134**, 404–18

Dumville, D N, 1977 Sub-Roman Britain: history and legend, *History*, **62**, 173–81

Frere, S S, 1974 *Britannia: a history of Roman Britain*, 2 edn

Hawkes, S C, 1974 Some recent finds of late Roman buckles, *Britannia*, **5**, 386–93

Hawkes, S C, & Dunning, G D, 1961 Soldiers and settlers in Britain, fourth to fifth century: with a catalogue of animal-ornamented buckles and related belt fittings, *Medieval Archaeol*, **5**, 1–70

———, & ———, 1964 Krieger und Siedler in Britannien während des 4 und 5 Jahrhunderts, *Bericht der Römisch-Germanischen Kommission*, **43–4**, 155–229

Johnson, S, 1976 *The Roman forts of the Saxon Shore*

Kent, J P C, 1951 Coin evidence and the evaluation of Hadrian's Wall, *Trans Cumberland Westmorland Antiq Archaeol Soc*, **51**, 4–15

MacMullen, R, 1963 *Soldier and citizen in the late Roman Empire*

Mann, J, 1977 *Duces* and *comites* in the 4th century, in *The Saxon Shore* (ed D E Johnston), CBA Res Rep, **18**, 1–15

———, 1979 Hadrian's Wall: the last phases, in *The end of Roman Britain* (ed P J Casey), Brit Archaeol Rep, 144–51

Marsden, E W, 1969 *Greek and Roman artillery: historical development*

———, 1971 *Greek and Roman artillery: technical treaties*

Richmond, I A, 1930 *The city walls of imperial Rome*

———, 1935 Trajan's army on Trajan's Column, *Pap Brit Sch Rome*, **13**, 1–40

———, 1940 The Romans in Redesdale, *History of Northumberland*, **15**

———, 1958 The fourth century AD and after, in *Roman and native in north Britain* (ed I A Richmond), 112–30

Schenk, D, 1930 Flavius Vegetius Renatus: die Quellen der Epitoma Rei Militaris, *Klio*, Beiheft 20

Stevens, C E, 1971 Progress report on Britain, in *Roman frontier studies 1967*, 34–7

Todd, M, 1978 *The walls of Rome*

Tomlin, R S O, 1976 *Notitia dignitatum omnium, tam civilium quam militarium*, in *Aspects of the Notitia Dignitatum* (eds R Goodburn & R Bartholomew), 189–210

Wacher, J, 1964 Cirencester 1963: fourth interim report, *Antiq J*, **44**, 9–18

———, 1975 *The towns of Roman Britain*

Webster, G, 1975 Small towns with defences, in *Small towns of Roman Britain* (eds W J Rodwell & R T Rowley), Brit Archaeol Rep, **15**, 53–66

Imperial campaigns and 4th century defences in Britain

John Casey

It is a commonly held view that archaeology and history in the Roman period have a compatibility, ie that physical phenomena in the archaeological record can be equated with otherwise attested historical events. In Britain in the 4th century there are a number of recorded events in good historical sources which, it might be expected, would leave some physical traces, not necessarily of destruction or disaster itself, but of the anticipation of and preparation to avert troubles, or *ex post facto* activity to prevent their repetition. The events in Britain which are well documented and might be expected to produce such categories of evidence are Constans' visit in the winter of 343, the expedition of Lupicinus at the behest of Julian in 360, the expedition of Count Theodosius in 368, and the intervention of Stilicho in *c* 398. To these four well attested interventions may be added a further three—two visits by Constantine the Great, the first in 312 and the second in 314 (Casey 1978), and an expedition by Magnus Maximus in, probably, 384 to conduct a campaign against the Picts (Casey 1980a).

Except in the case of Constantine's first visit, each of these events proceeded from barbarian attacks upon Britain and thus might be seen as situations in which the defences of the communities of Britain might have been improved where they already existed or provided where they did not exist. However, if we analyse the events outlined above, such evidence as we have points to intervention being largely confined to the northern frontier area. Ammianus certainly implies that Constans was in the north (xxviii, 3, 8); Constantine was reorganizing the frontier area; Lupicinus was sent because the Scotti were active and the Picts were ravaging the frontier area (Ammianus, xx, 1, 1), though it may be suspected that these troubles were less important than the need to get Lupicinus out of Gaul at a crucial moment in Julian's rise to power.

If these campaigns were in the north, this is not the case with the expedition mounted by Count Theodosius which, because of the involvement of the Count of the Saxon Shore (if this is the officer described by Ammianus), must have included military activity in the south of Britain, both in East Anglia and south of the Thames. Barbarian attack on towns is implicit in the account that we have of the episode and thus we might expect a response in terms of additions or amendments to the defences of the area. This expectation is reinforced by the light thrown on the activities of Theodosius by Ammianus' statement (xxviii, 3, 7) that after destroying the enemy he devoted himself to carrying out 'many necessary improvements', specified as 'restoring the cities and forts and protecting the frontiers with guards and outposts' (xxviii, 3, 1). We would naturally look to this circumstantial account of events as a basis on which to establish an archaeological dating horizon, and we should be able to assign structural phases in urban defences, forts, Hadrian's Wall, and the outpost forts of the northern frontier to this period. However, it is becoming ever more difficult to accept the account of events between 367 and 369 at face value. Notably, a survey of the numismatic evidence from the outpost forts of Hadrian's Wall indicates that they were abandoned long before the date claimed on behalf of Count Theodosius for their restoration (Casey 1980b). Indeed, these forts were almost certainly given up by Constantine during a reorganizing of the frontier which coincided with his visit to Britain in 312 to withdraw troops from Britain to participate in the civil war with Maxentius, this reorganization being the moment at which the office of *Dux Britanniarum* was created (Casey 1978).

If Theodosius had nothing to do with outpost forts, equally he may have had nothing to do with the creation of the system of signal towers on the Yorkshire coast. Arguments have been marshalled elsewhere which suggest that a reattribution of these works to Magnus Maximus may now be possible (Casey 1980a).

We are left then with the part played by Theodosius in the restoration of the cities of the diocese. In this area specific claims have been made which associate with the activities of Theodosius the addition of bastions (more properly to be called towers) to the walls of a number of towns. The case for this attribution has been most eloquently expressed by Frere (1974, 291) and has been reiterated by subsequent writers on the subject. Surveying the dating evidence in the light of the excavations at Great Casterton (Rutland), Frere writes:

> But it was the excavations at Great Casterton which threw more decisive light on the problem, for here two coins minted between 354 and 358 were found associated with two different bastions. Accordingly, they can hardly have been put up before 360; and, bearing in mind the limitations of archaeological dating-evidence and the scarcity of coins of the House of Valentinian in Britain before 369, we can hardly doubt that this reorganization of the town-defences of Britain was the work of Count Theodosius. He it was who recovered Britain in that year from the chaos and disorder of the great barbarian invasions of 367, and we have the record of Ammianus that he restored the towns.

This statement is based upon the coin evidence from Corder's excavation of Bastion II and Bastion III at Great Casterton in the 1950s (Corder 1961). In excavating Bastion III Corder found a coin closely dated to 354–6 in the filling of the ditch of the pre-bastion defences. This ditch was made redundant when the space which it occupied was needed for the foundations of the new series of towers; it was at the stage when the towers were erected that the ditch was backfilled. In Bastion II a coin of Constans was found in the construction level, which with the reverse type VICTORIAEDDAVGGQNN is datable to 346–8. An early copy of the post-354 FEL TEMP REPARATIO (Falling Horseman) coinage was found in a stratigraphically ambiguous position 'lost during or soon after the construction of the bastion' (Corder 1961).

In his report on the numismatic material from Great Casterton, Kent offered an admirable and succinct account of the coins found in relation to the numismatic history of the period of their issue and use. In this account Kent was at great pains to disassociate the coins at Great Casterton from a date as late as one which would bring the towers into the period of activity of Count Theodosius. This view is supported by the numismatic evidence from towers elsewhere in Britain and especially by the numismatic discoveries made during the excavations of a tower on the walls of Caerwent.

Other direct coin evidence in association with towers is not very great in quantity but it is consistent in composition. Two further sites are involved, Brough-on-Humber and London. At Brough-on-Humber a coin of Constantine dated to 330–7 was found in Bastion II in levels deposited against the town wall before the construction of the tower (Wacher 1969). In London Bastion 6 (Dukes Place), the footings of the tower were covered by a layer which contained a coin of Magnentius or Decentius (350–3), whilst the filling of the ditch of the city wall, a filling associated with the building of the tower, produced yet another VICTORIAEDDAVGGQNN coin of Constans of 346–8.[1]

Turning to Caerwent, the evidence for the early construction of the towers is eloquent. The excavation of the north-east corner tower in 1971 produced two results, firstly the dating of the tower itself by means of a coin hoard deposited during its construction and secondly the redating of the walls themselves by the identification of a debris level associated with the building of the walls and the relating of this level to a discovery made during earlier excavations.

To deal with the latter first. It has been clear for some time that the date assigned to the walls of Caerwent by Nash-Williams was too early (Nash-Williams 1964), and that ceramics found incorporated in the infilling between the early earthwork defences and the later wall pointed to a wall construction date late in the 3rd century at the earliest. In his excavations at Caerwent, Nash-Williams cleared the south-eastern corner tower and in the course of that work produced a coin of Constantinian date. This coin achieved considerable distinction in the archaeological record as providing a *terminus post quem* for the addition of towers to urban defences throughout Britain. It does indeed provide such a dating horizon at Caerwent, but more importantly it can be shown to provide a date for the construction of the walls themselves since the coin was stratified below a layer of mortar and sand deposited by the wall builders in the course of their work. This level was recognized on the line of the northern defences by the present writer during the course of his own excavations and seems to have been uniformly deposited around the circuit of the walls. The position of the coin may be summarized as follows:

A section through the tower revealed that a layer of sandy loam lay against the town wall and that this layer had been cut by the foundation of the tower. The layer was 'carried down to fill the foundation trench of the main wall' and contained an unworn coin of the Constantinian VRBS ROMA series dating to 330–5. Discussing the significance of the coin, the excavator commented that the texture of the layer which produced it was of such consistency that the coin might have worked its way down from a higher level. He concluded:

> The coin of *Vrbs Roma*, being outside the line of wall, may have dropped on the surface of the sand after the building of the wall, and worked its way down to the level where found. Whether this was the case or not there was nothing to show, but the mere possibility necessarily destroys its evidential value as far as the problem of the date of the wall is concerned (Nash-Williams 1930).

This judgement was made in the light of the misdating of the pottery from behind the wall. It does not reflect the excavator's original conclusion which is contained in correspondence between Nash-Williams and Fox now deposited in the archives of the Department of Archaeology of the National Museum of Wales. This correspondence, carried on during the excavation, includes a section drawing which makes clear that the coin was found at the base of the sand level. Nash-Williams writes:

> ... greatest news of all—we've bagged a coin in the bastion we're clearing as stratified as cld (sic) be—right at the bottom of a sandy layer (a rammed floor, as I suspect) *next to natural gravel and clay* (underlining in original text)...

Not merely is the sandy level represented as being of firm consistency, and thus unlikely to provide a medium through which the coin could drop, but at this stage the excavator thought that he was dating the tower. The problem of the wall date had not yet arisen. As was demonstrated in the excavation of 1971, the sandy layer was in fact the builders' deposits from the building of the wall and may be easily distinguished from the similar debris created in the erection of the towers. Since the level ran back into the wall foundation trench, a feature made possible by the fact that the curtain at Caerwent sometimes oversails its foundation, it follows that the walls themselves must date to after 330. An analysis of the series of Caerwent excavations demonstrates that the entire circuit of walls is of a single 4th century date and that the gates are of integral construction with the walls. The addition of towers followed shortly upon the completion of the curtain on the north and south walls with the traditional filling of the wall ditch to provide a firm foundation for the new work. A broad outer ditch was also provided. Inside the north-eastern corner tower the foundation trench was sealed by a working floor of rounded pebbles, masons' chippings, and mortar spills. In turn this was sealed by heaps of masons' chippings derived from work on the superstructure of the tower. On the working floor and sealed by the upper deposit of chippings was a coin hoard. This hoard, which must represent the coins carried by a workman engaged on the construction of the tower, consists of an extremely closely datable series of issues terminating with specimens of the earliest FEL TEMP REPARATIO coinage.

The FEL TEMP REPARATIO coinage was issued first in 348, probably in association with the celebration of the eleven hundredth anniversary of the foundation of the city of Rome. Coins were issued in three denominations and the first moves in the reform of the currency, which the new types represent, were made in the west by Constans. The first issue of the reform was of intermediate denomination with the reverse type depicting a soldier leading a barbarian out of a hut under a tree. Sharing mint marks with this issue is the smallest denomination of the series with the reverse type of a phoenix standing either on a globe or a pyre. The third component of the reform was the highest denomination which was issued after an interval of short, but uncertain, length from the fractions. The high denomination consisted, essentially, of two types, one depicting the emperor standing in a galley steered by Victory and the other showing a fallen enemy horseman being speared by a Roman soldier. In the western provinces these coin types were superseded by issues of Magnentius between 350 and 353. On the resumption of imperial power in the west only a modified version of the Falling Horseman type was issued.

In the Caerwent hoard the terminal date is given by the presence of a single specimen of the Hut/Barbarian issue and three of the Phoenix/Pyre coins; there are no coins of the Galley/Victory or early Falling Horseman type in the hoard. There is no representation of Magnentius. A terminal date for the hoard of late 348 or 349 is therefore indicated. The condition of the terminal coins would be consonant with a deposit date coincident with the issue date of the coins. The composition of the hoard as a whole has a bearing on the dating offered here and reflects on the general problem of the date of coins associated with towers elsewhere in Britain.

The VICTORIAEDDAVGGQNN coinage has been dated to the period 341–46 (Carson *et al* 1960) but recent research suggests that the true date is later than this. It is now recognized that the FEL TEMP REPARATIO reform took place in 348 and that there was a hiatus in the issue of coinage between, it is suggested, 341 and 346 (Kent, pers comm). At the latter date the VICTORIAEDDAVGGQNN coins came into circulation for the first time and were issued in quantity until superseded by the FEL TEMP REPARATIO issues. The effect of redating the VICTORIAEDDAVGGQNN coinage to 346 is, of course, to distance the deposits of such coins in towers and wall ditches some way from the visit of Constans in 343.

If we recapitulate the evidence assembled thus far, we have a consistent series of deposits containing VICTORIAEDDAVGQNN coins, or earlier Constantinian issues, in the filling of redundant wall ditches or construction levels. We also find an unambiguously stratified coin of 354–6 in Bastion III at Great Casterton and the closely dated hoard of 348–9 at Caerwent. There is only a single instance of a copy of the FEL TEMP REPARATIO coinage of post-354 type, and that a very early copy, in an ambiguous stratigraphic position at Great Casterton. Yet it is copies of the FEL TEMP REPARATIO coinage of post-354 type that fill the hiatus in currency issues which was noticed by Frere. The origins of these copies lay in a rescript issued by Constantius II at Arles in 354 (*Codex Theodosianus*, 9, 23, 1). Probably intended only to demonetize the coinage of the usurper Magnentius, the rescript was couched in terms of such ambiguity coupled with condign threat that it effectively operated on the still circulating issues of Constantine and his sons of the period 330–48. Whatever the intention of the law, the effect was to drive out of large-scale circulation pre-FEL TEMP REPARATIO issues and to produce a huge volume of unofficial copies of the post-354 Falling Horseman coinage. Official issues of this coinage were themselves in extremely short supply in the west, an effect of Constantius' distrust of his Caesar Julian, a distrust which found expression in a policy of preventing Julian having access to coinage with which to pay his troops (Ammianus, xx, 8). As a policy to prevent usurpation this proved ineffective; as a hindrance to commerce it enjoyed widespread success. Julian's removal of the army from Gaul to campaign against Constantius, the subsequent Persian war, an abortive coin reform, and two rapid changes of dynasty all contributed to keeping the Prefecture of the Gauls short of coinage between 354 and 364. The very rapid proliferation of copies may be judged by hoard evidence. Although hoarders might be expected to favour regular coins and specimens of higher intrinsic value, conditions occasionally forced them to mix the contents of their hoards. Just such a hoard is that from Oldcroft (Glos). Concealed between 354 and 359, and perhaps nearer the former date than the latter, of 3258 coins examined no less than 1397 (42·8%) were post-354 FEL TEMP REPARATIO copies and only 223 coins (6·8%) were pre-348 issues.

The cumulative coin evidence is slight on a numerical basis but, on the evidence of the Caerwent hoard, there is nothing in it which warrants the association of the addition of towers to the walls of the towns of Roman Britain with the period of Count Theodosius's activity in the diocese. On present evidence it also seems unlikely that the visit of Constans had any discernible effect. As far as the numismatic bracket, outlined in this paper, is concerned an inception of work at, or just before, the accession of Magnentius is indicated. Possibly the removal of forces towards the Balkans gave impetus to this policy.

One cannot ascribe this policy, if such it was, to imperial intervention. The provision of towers to the defences of a number of towns, the provision of new circuits of wall at Caerwent, Ilchester,[2] Thorpe, Catterick, and Great Chesterford, and the completion of the defences of London by the construction of a riverside wall must be seen in the light of imperial financial regulation at the period of their construction. It is very unlikely than any imperial financial intervention, in the positive sense, contributed to the new works. Indeed the contrary condition may well have prevailed, since the accession of Magnentius will certainly have cost the cities a contribution of 'crown gold'. Defences were a matter for communities rather than for imperial governments and their agents and it was from the funds of individual towns that such grandiose schemes as that of Caerwent were funded. Before the reign of Constantine, towns had access to money from a number of sources, notably the revenues from leases of town lands, from custom duties, and from the fees paid by town councillors and magistrates on taking office. All of these revenues were confiscated by Constantine and, although briefly refunded by Julian, again reverted to the imperial fisc in the reign of Valentinian who relinquished only one-third of them to the cities for use in the restoration and construction of public works. In effect, just at the time when large-scale mural works were being carried out in Britain, towns should have been at a low financial ebb. Obviously the picture of urban decay that is now current needs modification. In the light of the financial background of the empire it is perhaps possible to measure the wealth of individual towns, or the size of the benefactions which their citizens could still afford to bestow upon them, by the nature of their additional defences, relative poverty being indicated by the provision of occasional towers and the use of secondhand materials, and wealth, or relative wealth, shown in the use of new material and superior technology such as hollow towers and whole new wall circuits.

It was not until after the date proposed in this paper for the renewal of urban defences in Britain that the Roman state took any practical steps to assist in the provision of funds for public works. Apparently alarmed at the dereliction of cities Valentinian remitted one-third of the annually confiscated city revenues specifically for the repair and commission of public works. This concession was continued by his successors and a series of rescripts preserved in the Theodosian Code indicate that imperial concern grew with the passage of the second half of the 4th century. It would appear that the towns of Britain had anticipated this concern by some decades at least.

Notes

1 Information quoted by the kindness of Dr Hugh Chapman and Mr John Maloney of the Museum of London and the Department of Urban Archaeology.

2 Authors unpublished excavations.

References

Carson, R A G, Hill, P V, & Kent, J P C, 1960 *Late Roman bronze coinage*

Casey, P J, 1978 Constantine the Great in Britain—the evidence of the coinage at the London mint, AD 312–14, in *Collectanea Londiniensia: studies in London archaeology and history presented to Ralph Merrifield* (eds J Bird *et al*), 181–93

———, 1980a Magnus Maximus in Britain: a reappraisal, in *The end of Roman Britain* (ed P J Casey), Brit Archaeol Rep, 66–79

———,1980b The coins from the excavations at High Rochester, 1854 and 1856, *Archaeol Aeliana*, 5 ser, **8,** 75–87

Corder, P, 1961 *The Roman town and villa at Great Casterton, Rutland. Third report*

Frere, S S, 1974 *Britannia: a history of Roman Britain,* 2 edn

Nash-Williams, V E, 1930 Further excavations at Caerwent, Monmouthshire, 1923–5, *Archaeologia,* **80,** 229–88

———, 1964 The Roman town of Venta Silurum and its defences, in *Carnuntina* (ed E Swoboda)

Rhodes, J F, 1974 The Oldcroft (1971–2) hoard of bronze coins and silver objects, *Numis Chron,* 7 ser, **14,** 65–74

Wacher, J, 1969 *Excavations at Brough-on-Humber, 1958–61*

PART III MILITARY ARCHITECTURE

The construction of urban defences

R M Butler

The study of the construction of town defences is not generally undertaken for its own sake, because, apart from a few monumental gateways or unusual towers, they are not artistic or aesthetic achievements, varied with decoration, nor interesting as dramatic feats of engineering, as are aqueducts or even sewers. It is pursued in the hope of determining the date of these features and of distinguishing different periods of construction. Both antiquaries in noting down descriptions of standing ruins and excavators in observing sections of foundation have usually recorded aspects of construction in considerable detail, in the hope that comparison of the variable features will enable them or their successors to formulate some general theory of dating. By this method it is thought that any defence work discovered can be fitted into a picture of the response either of central government or of local communities to known threats from barbarian invasions, revolts, and civil wars. Perhaps our hopes of fitting everything neatly into general theories of reaction to threats or disasters, of imperial initiatives, of general patterns in the growth and decay of urban communities, even in one province, may be pitched too high. We can look back over many promising theories about Roman Britain, from Agricola's campaigns and the sequence of building on Hadrian's Wall to what really happened in the 5th century. So many of these have had to be modified or discarded over the past few decades that in a few years' time any schemes put forward at this conference may be regarded with pity, or even by harsher or younger judges with contempt.

This paper can therefore only be a reminder, a recapitulation, a check list of the variable features observed in Roman town defences in Britain and Gaul—I claim no wider detailed knowledge—which every investigator must consider in comparing his discovery of a length of curtain wall, a section of foundation, or a fragmentary tower with those found elsewhere.

The plans of urban defences, the size of the area enclosed, the extent to which earlier buildings, like the amphitheatres at Amiens and Tours and the forum at Bavai, were incorporated or respected, or whether later structures, like the aqueduct at Fréjus, encroach on walls, depend so much on local conditions that they cannot really be used in building up a general pattern of dating. Rectilinear and polygonal plans of wall circuits are as common in the Early Empire as later; though we may expect that on an unencumbered site Roman military architects would have preferred a rectangular plan, they seldom had the task of defending a simple rectangle of dwellings. Winding rivers and sprawling agglomerations of buildings complicated their task. The number of monuments left standing to be incorporated or ignored also varied and we can only guess at the motives which left such massive works as the great tower temple of Vesunna at Périgueux or the Piliers de Tutele at Bordeaux just outside the defences of those cities.

The variable features to which I refer are the nature of the foundations employed, the use of timber, whether as piles supporting footings or as lacing or load-spreading beams in the core, the characteristics of that core, and the facing, maybe with tile-courses, small ashlar, large blocks, well-coursed, or with patterns in different coloured or shaped stones as decoration. Since parapets so rarely survive any trace of them is a bonus, not a feature found more than once in 50 defence works. The thicknesses of footings and curtains vary according to whether the walls were free-standing or added to pre-existing earth ramparts, though unless a number of observations at widely separated parts of a circuit agree, it cannot be known whether one section is normal or not. Towers too—I call them towers since Roman writers described them as *turres* and the use of the word 'bastion' is misleading in view of its technical meaning in renaissance and later fortifications—come in many possible forms. They can be internal only, rectangular, square, or semicircular, external in the same shapes, plus U-formed, horseshoe, half-hexagonal, half-decagonal, or, more rarely, fan-shaped; they can bestride the walls as circles or rectangles. Where more than the lower few feet survive the upper part can turn out to be hollow on a solid base. If more survives still, up to two upper storeys with windows and doorways may be found. Gateways come in many forms and are comparatively less frequent survivors, with four or two passageways, or, most commonly, one, with round, hexagonal, or rectangular inner courtyards, and with flanking towers of all shapes. Narrow posterns near or through towers, drains large and small, may also occur.

General syntheses for different countries or provinces are available (see bibliography) and the studies of particular sites which have been summarized in this volume happily make it unnecessary to go through examples. However, in spite of so many discoveries in the past ten years, the old distinctions between British and Gallic town defences still, I believe, remain. It may be that the more thorough investigations at such places as Gloucester, Lincoln, and London, which have shown that the history of their defences is even more complicated than was suspected, have made the apparently more uniform picture in Gaul more doubtful. The situation there has not apparently altered greatly over recent years, perhaps because new discoveries have actually done no more than confirm the old observations and assumptions, or because the evidence which would show that earth ramparts once existed,[1] that projecting towers were additions to earlier walls with none, that several phases disturb the satisfactory unitary scheme, has not been looked for and so not found.

Although the 400 years from 1100 to 1500 were perhaps more disturbed by external dangers and civil wars than the four centuries of Roman rule, an example from the field of medieval town defences may suggest that as varied a pattern could be expected in the earlier period. At York earth ramparts with timber gates (themselves concealing the stone walls of the Roman legionary fortress and *colonia*) were first equipped with stone gateways, then had stone walls built on top, then two varieties of towers rounded and square, loftier gatetowers, added polygonal towers, barbicans before the gates, and finally additional rectangular towers (RCHM 1972). Should as complicated a history within a comparable length of time perhaps be envisaged for many of the defences we are studying?

Contemporary with the town defences and often not far away from them are the defences of forts. Attempts to classify the later forts by plan, walling, tower shapes, and gateways have met with varying success. Even in our own fairly similar Saxon Shore forts there seem to be four periods within a short time, while a different four can be traced on and behind the Rhine frontier, including at each stage forts of very different design, even if they show the common characteristics of irregular or little known internal buildings, thick walls, and projecting towers. We can encounter forts as different in design as Kaiseraugst and Altrip, as Pachten and Irgenhausen, as Pevensey and Alzey, while later legionary fortress defences can vary, as does York with its white limestone walls relieved by broad tile-courses and eight massive polygonal towers along the river front, or Strasbourg with timber-laced semicircular towers added in front of the closely-spaced square internal turrets and curtain walls thickened from 0·9 m to 3·7 m (RCHM 1962, 6–37; Forrer 1927, 39ff; *Gallia*, **32** (1974), 381–3; **34** (1976), 392–3).

Similarly, city walls can be very different in detail. Take Toulouse, with 3 km of landward wall enclosing 90 ha (Labrousse 1968, 237–50). Footings were of river pebbles set in hard mortar carrying a similar core faced at the base with limestone ashlar and triple tile-courses, the whole 2·4 m (8 Roman ft) thick. Above the facing was entirely of tile—Ausonius's *coctilibus muris*—and the towers were circular and hollow, 10 m in diameter, and of one build with the walls (except for three larger examples with rounded or polygonal fronts). The north gate (La Porterie), the only one of the three for which details are known, had projecting semicircular fronted towers, three passageways, a round internal court, as at Avenches, and was decorated on its outer facade with attached Corinthian columns. The river wall was later with reused material at the base, cross walls of tile dividing the core into compartments filled with tile and pebble concrete, and tile facing. No towers and only one postern are known from this side. With some differences, the extensive defences of Tongres and Trier are on a similar pattern, all three apparently fortified *c* AD 200–30.

Such defences are plainly very unlike those of an average British town—if there can be such a place. Canterbury, for instance, covered 48 ha with a rampart 7·6 m wide and 2 m high and a wall 2·3 m thick of coursed flints set in yellow mortar and 6 m high, if the parapet which we have twice seen was Roman (Wacher 1975, 188–9; *Britannia*, **8** (1977), 424; **9** (1978), 468). Towers were internal and square—the externally projecting U-shaped towers all seem to be medieval additions—and most of the six gates had narrow passages under tile arches. Chalk quarries near the walls and tile kilns close by could supply the material used, whereas some of that at Toulouse had to come from 70 km away. A date in the 270s is apparently still assigned to the walls of Canterbury. Both its defences and those of the city on the Garonne contrast sharply with those of Le Mans, Périgueux, Tours, and Carcassonne or with the group Beauvais-Senlis-Soissons in Gallia Belgica studied by Johnson (1973).

Details which we might expect to encounter normally in Roman town walls may now be briefly surveyed.

Foundations

These are usually so strong that they have borne for centuries the weight of curtain wall and towers even when they have been cut about or undermined by cellars, pierced by windows and chimney shafts, when higher medieval walls have been piled upon them, or houses leant against them. Only rarely do we find a collapse, as at London or Scarponne (*Gallia*, **30** (1972), 349–54), undercut by a river at Orléans into an earlier ditch (*Gallia*, **34** (1976), 325–6), or in the fort at Lympne, caused by a landslip. It has been claimed that in southern Gaul the earlier town walls have little or no foundations, sometimes resting directly on natural rock, but this does not seem to be the case normally. The footings may occasionally be supported on piles, but such laborious underpinning is, as one might expect, usually confined to marshy or unstable sites, of which the London river front is an example. Other instances are at Gloucester, Rennes, Tongres, and Xanten. A rarer technique is a sort of timber framing or lacing in the foundations or lower part of the core, whether to spread the load of the superstructure on unstable ground or to tie together towers and walls. The clearest instances are, however, on military sites, as in the thickening and towers added at Strasbourg, in the forts, as they seem to be rather than towns, of Andernach and Saverne, and in the English Saxon Shore forts at Pevensey, Portchester, and Richborough, two at least probably built after 350 and all work done after AD 250. The odd length of timbering in the London river wall appears to be something of this sort, and the foundations at Dax sound as if they resembled those of Pevensey. The technique was used more widely in the signal towers of the Rhine frontier in Switzerland, attributed to Valentinian I. It reappears again in England on a large scale in the massive foundations of the Norman York Minster and the contemporary Richmond Castle. Rough arches to spread the weight of the city wall, as used in medieval times at Southampton Castle, were made at Trier, where the amphitheatre was incorporated in the eastern defences (Wightman 1970, 90–8).

Some, probably most, footings were carefully laid in a foundation trench, usually in Britain starting with a thick layer of cobbles set in clay and in Gaul with similar cobbles set in mortar, though angular pointed stones forced into the subsoil have been reported at Dijon and London. The great difference which distinguishes most of the town walls of Gaul from nearly all those in Britain is of course the reuse of previously worked blocks, from civic buildings, cemeteries, and shrines. Since this is covered by Blagg's paper in this volume, I will only touch on it briefly. This practice is found in forts of the later period, presumably contemporary with the town defences, but quantities of reused material are smaller, no doubt because the available material was much less rather than because there was some reluctance to use it. At Alzey, within the fort of the 360s, an internal building, probably of timber, stood on footings made from altars and dedications derived from a shrine to Apollo (Behn 1929/30; 1933). At three of the Saxon Shore forts, notably Richborough, and in the north wall of Chester, which incorporates many tombstones, this reuse is found. Road forts or small towns, like Arlon and Neumagen, have yielded fine reliefs, but nothing to compare in quantity with the masses of material from Bordeaux or Sens. Recent discoveries of the sort of inscriptions and sculptures which excited 19th century antiquaries are at Metz (seventeen fine funerary *stelae* in a length of 15 m of curtain wall (*Gallia*, **34** (1976) 363–7), Rennes, with dedications and statue bases from a temple of Mars Mullus (Bousquet 1971), and Bourges, where a building within the late defences was similarly constructed on a base of large reused blocks (*Gallia*, **36** (1978), 262). Among British town defences only a few places have anything comparable: Kenchester, apparently with a milestone to compare with the ten used to make a postern at Rennes or the four at Bitterne, possibly Bath, if not medieval, the lower west gate at Lincoln, and most notably London, with its river wall and eastern series of added towers. The height to which this reused material was used, normally apparently depending solely on its weight and careful fitting together, rather than on mortar or on metal clamps, found in the Hellenistic walls of Marseilles, to hold the pieces firm, varies according to its availability—7 m at Périgueux, 6 m at Bordeaux, at Bourges in places, and at

Verona in Italy, but only 1·5 m at Le Mans and less at Metz and Nantes.

The core

The core of all Roman town walls, both in Britain and in Gaul, seems to be described in much the same way and to look very similar: mortared rubble or rubble concrete was cemented together to make a solid mass, strongly resisting destruction and decay. Sometimes so-called herringbone work or counterpitched courses have been noted—they are recorded at some ten places in Britain from York to Dorchester, and at Metz and Toul and in several of the late forts. Since this is a method of building with irregular-shaped stones used at any time, it cannot serve as a dating criterion, even though apparently most common in the early Norman period. Cross walls dividing the core of the walls into compartments, whether for strength or due to the apportioning of the work to separate gangs, have been noted at Toulouse and perhaps at Rennes, but are otherwise most unusual. Mortars of different composition, appearing whiter or creamier or redder, can be used to distinguish work of different periods on the same site, as in the two walls of the fortlets at Bavai and Famars (Will 1963). However, no general rule for differentiating mortars of different centuries seems to be accepted. Perhaps measurements of remanent magnetism may some day help in this. However, in view of the vast quantities of mortar required on a circuit, probably varying in mix week by week as supplies or supervisors varied, and considering the number of repairs to defences carried out through the 1500 or more years since their erection, only a large number of samples from all parts of a town's circuit, careful chosen and analysed, could reveal any consistency in the mortar used.

The facing

The facing of the walls, often rough and carelessly finished where, as in many cases in Britain, it was to be concealed by a rampart, seems usually to have been unrelieved by any decoration. A chamfered plinth at the base can appear both in the 2nd and 3rd centuries, as at Exeter, Gloucester, and Tongres, or in the 1st, as at Vienne. Occasionally, walls are preserved to a height which reveals that there was a string course below the parapet, at Verona for example. Decoration by patterns of different coloured stones in the facing may have been more common than the few examples preserved suggest; so few towns retain a long enough stretch of wall facing to be sure. As it is, Cologne, with its remarkable angle tower, and Fréjus in the 1st century, and Le Mans, Jublains, Périgueux, and Rennes in the late 3rd century have this feature. The latter group, with patterns of triangles, hourglass-shapes, lozenges, and chequers, rather like those in 17th century ornamental brickwork but carried out in square blocks of brown sandstone or grey limestone, have close resemblances to decoration on the walls of the Saxon Shore forts of Pevensey and Richborough.

Neat, regular, plain, coursed stonework can be found from the 1st century to the 4th and is not in itself a criterion of date. Tile-courses, too, though almost ubiquitous in the Gallic town defences assigned to the late 3rd century, need not mean a late date for the work in which they occur. In Britain the defences of Colchester are apparently of the 1st century, while those of Caistor-by-Norwich, Verulamium, and London are late 2nd or early 3rd in date. At Silchester and Dorchester, Dorset, flat stone slabs serve the same purpose in the same period. Various combinations of tile-courses have been observed, with single, double, or, most commonly, triple bands, differently spaced in relation to the stone-courses. Some 90 variants were counted at Carcassone, with 68 being single courses (Poux 1922, 117 ff). The most usual spacing of these tile-courses, at about 1–1·5 m intervals, seems to be related to the stages in which the work was executed. Before the planking on the scaffolding from which the facing was built was moved up, layers of tiles served to level off the uppermost facing-course and to bond it with the rubble concrete core, which had been laid or poured between the outer and inner faces. (Although observers at two towns have claimed that the facing was applied to an earlier core, this technique, used in modern concrete structures, seems to be unlikely and unsupported by good evidence in the defences we are examining.) In several cases, for instance Beauvais and London, some tile-courses run right through the thickness of the walls, but they usually only penetrate to a depth of one or two tiles, and the courses on the inner and outer faces need not match.

Whether the proportion of tile-courses to stone-courses means more than the varying availability of local supplies or can be used to distinguish local fashions in wall construction is not clear. Certainly at Dax and Toulouse there were far more tile-courses; at the latter city the upper part of the walls was entirely faced in tiles, providing a resemblance to the later tile-faced walls of Rome and most of the circuit at Salonica. The evidence for dating provided by tile stamps has not been much used in Britain and Gaul, or is not forthcoming, though where, as on the Rhine frontier, they have been examined in detail, it seems that stamped tiles of a unit, an official, or a company, may have continued in use over quite a long period with little variation in the stamp, probably because old stocks were being used. Fortunately modern forgeries do not normally complicate matters, as on the Saxon Shore at Pevensey and Lympne (Peacock 1967).

Where good lengths of facing are preserved, putlog holes can be recognized at about 1·5 m intervals, both vertically and horizontally. There may be some significance in their placing in or just above tile-courses, but these may be just local variations, and different positions for these holes, apparently usually left empty, may occur around the same circuit of defences. Careful recording and comparison is needed before any general deductions can be drawn from their varying locations. For only a few Gallic town walls has the construction been as carefully examined as at Rennes (Merlat 1958). Masons' or quarry marks on dressed stonework, found on blocks in the Porta Nigra at Trier, are rarely observed. Scatters of chips from masons' working, as at Aldborough, may reveal how material was dressed on the spot and spreads of mortar which can be associated with the original building of walls sometimes seal or contain dating evidence.

Changes in the facing or core can reveal the work of gangs constructing short lengths. These are seldom as obvious as at Pevensey and Richborough, where such changes occur at intervals as small as 15–20 m, but have been observed at London, Verulamium, Nantes, and Rennes. Breaks in the core concealed by the facing only appear when a good length can be examined closely, perhaps in the course of demolition: Metz and Caerwent showed this feature.

Although arrow slits in Roman city walls do occur, as in the added upper part of Rome's or in the outer wall at Constantinople, the only example reported in Gaul is at Dax, where Roach Smith (1861) saw horizontal slits before the demolition of most of the circuit. The buttresses at Xanten could possibly have supported an arcade carrying the wall-walk with openings through the thinner curtain wall, as in the medieval walls of Norwich and Yarmouth. Apart from gates and posterns, the only apertures through curtain walls are drains, which perhaps should have been mentioned under footings, since they would have had to be allowed for in building that stage. In Britain they have been recorded in the *colonia* walls at Colchester and somewhat later at Silchester, and in Gaul in the defences of Carcassonne, Le Mans, St Bertrand, and Tours. Hellenkemper showed how they were catered for at Cologne.

Traces of undoubted Roman parapets with merlons and embrasures survive so rarely in the area I am considering that they can be discounted as dating evidence, and to see examples one must visit Rome, Barcelona, or Salonica. However, worked stones apparently capping battlements have been found at Nîmes, Trier, Verona, Towcester, and probably now at London, and moulded tiles from parapets at Grand. The surviving parapet at Pevensey and that preserved in the church wall of St Mary, Northgate, Canterbury, may also be Roman (*Britannia*, **9** (1978), 468; Tatton-Brown 1978, 80–1).

Towers

It is from the plans of towers that most attempts at collecting criteria for dating have hoped to gain more certainty. In Britain recorded towers on wall-circuits are still very few when compared with the 30 or more found around a medium-sized Gallic walled town or the 400 projecting from the walls of Rome. London has the most with 20+ known, followed by Caerwent with 11. On town walls this side of the Channel only about 30 internal and 80 external towers are so far known, many from mere fragments. Although there are so many possible variant shapes, those usually found in Britain amount to four:

1 Internal, square or rectangular, and hollow are found at at least seven towns, normally where the wall has been added to a pre-existing rampart;

2 External and rectangular, with the best instance at Brough-on-Humber, which may well be a military site;

3 External and D-shaped, solid or hollow, as at Caistor-by-Norwich, London, and Verulamium;

4 more rarely, polygonal, with examples at Caerwent, Cirencester, and perhaps Kenchester.

The odd fan-shaped towers at Ancaster and one somewhat similar at Aldborough may be counted variants of the D-shape. Whereas most of the rectangular internal towers are of the same build as the walls and are presumably derived from those of forts, the others have nearly all been added at some time in the 4th century—*c* 370 seems to be the date most favoured, but this may soon have to be revised again—or, where there is some doubt, can probably be shown to be additions.

In Gaul and Germany the situation is very different. Of the twenty or so towns fortified before AD 200 those at all well known had hollow round towers astride the curtain wall. Circular towers are also apparently the norm on the later town or road station defences in the east, with both solid and hollow ones occurring at the same site. In the west D-shaped towers are usual, mostly just projecting externally but some, like the group in Belgica distinguished by Johnson, projecting slightly with rectangular backs within the curtain. Other variants seem to be found in the frontier forts, though there are defences, like Neumagen and Arlon, small enough to be called forts, or others, like Andernach and Koblenz, of the size of small towns, which share with a dozen or so road stations the features of round towers and gateways under a square tower on rectangular supports.

While in Britain only the towers of Caerwent are preserved to a height great enough to distinguish internal floors, and windows in towers only survive at the military sites of Pevensey and York, several cities in Gaul retain towers with one or even two storeys standing above the parapet walk. Le Mans and Senlis are the best known, and on them I based the reconstruction drawing seen so frequently at this conference (Butler 1959, 39, fig 5), but Bourges and Carcassonne also have rooms with three tile-arched windows in the outer face and doorways to the wall-walk. Since the usual height of the walls is about 6 m or 20–25 ft, and these rooms were 3 m or more high, the original height of towers must have been 15–20 m (50 or 60 ft). The better preserved towers farther afield also rose above the parapet of the walls; presumably this was normal. We can only speculate whether they once had flat roofs or conical tiled, or even lead-covered, tops. Mosaics, coins, and manuscript illustrations show both terminations. The use of artillery from such towers would affect the size of windows and the roofing needed, but this subject is dealt with by Baatz.

Gateways

On gateways I hesitate to say anything for the topic has been amply debated, the examples in Gaul and Britain are comparatively few, and each one is somewhat different from the others. Those with more than one passageway, usually four, are less common, with ten examples in Britain, twelve in Gaul, three or four in Germany, and two in Switzerland. Of these only the Porta Nigra at Trier, the two gates at Autun, and battered fragments at Colchester, Lincoln, Arles, Fréjus, Nîmes, and Vienne, rise above foundation level. Even if we descend the scale to the commoner class of gateway, a single arch between flanking projecting towers, only Caerwent and Die have structure, Lincoln, Nantes, and Périgueux foundations. Posterns are more frequent survivals, as at Le Mans or Sens, and remained long enough at Bourges, Dax, and Carcassonne to be recorded. In the smaller road stations nearer the Rhine, like Andernach, Bitburg, and Junkerath, they emerge through hollow round towers, much as they do through towers of other forms at Richborough and Pevensey. Since only at Lincoln, where five gates are known in detail (Thompson 1973), is there any assurance of knowing anything like the complete plan and elevation of more than half the gates of a Roman town in Britain or Gaul—and gates are a feature more likely to be destroyed or modified than towers—there is much to be discovered. However, there is good hope for future progress, for, even if a standing gateway is unlikely to be found concealed in other buildings, plans can be revealed by excavation, as recent discoveries at Lincoln, Cirencester, and Toulouse show. The incorporation of earlier arches, symbolic and decorative entrances to cities rather than defensive obstacles, in later defences, as at Colchester and Rheims, may be found to have occurred more frequently than is suspected.

The question whether military fortifications influenced town defences or whether the example of earlier town walls changed the form of forts should allow of a provisional answer in view of the large number of examples of both. In nearly all British towns an earth rampart is the first form of defence, or it accompanies stone walls. When these walls were added the internal towers recall those of forts. Gateways flanked by rectangular towers or with a single tower over a passageway are again reminiscent of those of forts. Yet lofty projecting towers, both at entrances to cities and along their curtain walls, existed long before they were used on forts. Probably the stone gateways of the legionary fortresses, like the late north gate at Regensburg, looked very like city gates before the dangers of the 3rd century compelled the widespread strengthening of both civilian and military fortifications. From the survey of the north-western frontier zone made by von Petrikovits (1971), it looks as if the thicker walls and externally projecting towers came to the military from the civilian sphere, but possibly the designers of the town defences were in imperial service and methods tried out on a large scale in the interior of Gaul were applied on the frontier a little later as the forts were gradually modernized and new ones built.

The various 3rd century emperors known to have initiated town defences or thought to have been responsible for them—Aurelian, Probus, Diocletian, Constantius, perhaps Postumus and Carausius—are only separated by a few years. Pottery experts must tell us whether debris sealed by a wall or dropped by its builders can be confidently assigned to the six-year reign of Aurelian, rather than to one of his immediate successors, or to Diocletian, only ten years later. The only inscriptions dating town defences in Gaul and Germany are of Augustus at Nîmes, of Diocletian and Maximian at Grenoble, of a 4th century praetorian prefect at Narbonne, and possibly of Gallienus at Cologne. Coin finds, unless forming a sealed deposit, as at Amiens, ending in 252, are often doubtful or equivocal. We must still in Gaul turn to the latest date occurring in inscriptions reused in the foundations (provided we can be sure that there is no possibility of a later repair) and in Britain work the harder way from pottery evidence. One day perhaps scientific dating methods such as thermoluminescence, residual magnetism, and tree-ring dating may be sufficiently developed to give certain dates to any structure, including fortifications.

Without much unexpected new evidence we are never likely to know who built the later town walls. At Rome in 273 the city guilds were pressed into service, while 25 years earlier the military built the circuit of walls around the colony of Romula (Resca-Dobrosloveni) just before the Goths invaded Moesia (Dessau 510; MacKendrick 1975). Normally, even if an emperor claims or is given credit for a defence work, the initiative must have come from lower down the hierarchy. Just as William the Conqueror or Oliver Cromwell are popularly praised or blamed for much of which they never dreamt, so Aurelian and Constantine may never have visited Dijon and Tours, where walls are attributed to them by much later sources. Decisions to fortify the towns of a province or a diocese were presumably taken at praetorian prefecture level rather than in the imperial consistory; they have consequently not found their way into collections of laws or edicts, as has the decision of 396 to use debris from demolished pagan temples for city defences. If landowners and city councils were effective pressure groups, influencing the central government in the ways Jones describes, they may have persuaded it to pay for defences needed for military purposes to shelter the mobile units of the field army. Those forces which were quartered in the cities, with all the disputes about provision of heating, free meals, baths for higher officers, and other perks which the law codes reveal (Jones 1964, 631–2), would no doubt have felt even more comfortable within defences—to the advantage both of the government and of the citizens, though they did not keep out 'friendly' troops, who could be almost as rapacious and destructive as enemy invaders, especially when most of the Roman army in the west was of Germanic stock. If peasants bringing produce to market were forced to carry out builders' debris from public buildings, as at Antioch (Jones 1964, 736), and decurions were expected to pay for such work as a civic duty, even though they no doubt managed to shift most of the burden to *humiliores*, the government may have been able to build defences cheaply and to take the credit. In 374 Valentinian and Valens had to refund to the cities rents from confiscated estates to enable them to maintain their walls and other public buildings (Jones 1964, 737). In 396 Honorius directed that provincial governors should see that the councils and citizens built new defences or repaired old ones; all material coming from the demolition of temples, where worship had been forbidden for the past few years, was to be used for the walls (*Codex Theodosii*, 15.1, *De operibus publicis*, lex 34 ff). Of course by this time all the British and Gallic walls had long since been built—or so we suppose—but this instruction suggests the sort of provisions which lesser officials might have issued on previous occasions for the use of tombstones or pagan monuments or material from derelict buildings.

If I have a conclusion it is that every town should be treated as a separate entity whose history may be as different from its neighbours as Leicester's from Nottingham's or Stamford's from Peterborough's. We should by all means try to discern or construct patterns or models but be prepared to discard them readily, not cling on to them for years. Although as yet it still seems possible to retain the pattern for Gallic town defences which seemed valid twenty years ago, I am ready to reject it for another, if towers are shown to be additions in many places, if new and conclusive dating evidence appears, or if—most unlikely possibility of all—an inscription or two reveal a builder's name. This study is full of surprises. Any study based largely on archaeological evidence must always be so.

Note

1 A low rampart, perhaps only material from a foundation trench, appears to have existed at Beauvais.

References

Behn, F, 1929/30 Ausgrabungen in Kastell Alzey, *Mainzer Zeitschrift*, **24/5**, 71–99

———, 1933 Neue Ausgrabungen in Alzey, Zweiter Bericht, *ibid*, **28**, 43–59

Blanchet, A, 1907 *Les enceintes romaines de la Gaule*, Paris (repr 1979)

Bousquet, J, 1971 Inscriptions de Rennes, *Gallia*, **29**, 109–22

Butler, R M, 1959 Late Roman town walls in Gaul, *Archaeol J*, **116**, 25–50

Forrer, R, 1927 *Strasbourg-Argentorate*, **1**, *das roemische Strassburg–Argentorate*

Grenier, A, 1931 *Manuel d'archéologie Gallo-Romaine*, **1**, Paris

Johnson, J S, 1973 A group of late Roman city walls in Gallia Belgica, *Britannia*, **4**, 210–23

Jones, A H M, 1964 *The later Roman Empire, 284–602*

Labrousse, M, 1968 *Toulouse antique*

MacKendrick, P, 1975 *The Dacian stones speak*

Merlat, P, 1958 Rapport sur la portion du mur d'enceinte gallo-romain de Rennes découverte, 18 Quay Duguay-Trouin, *Annales de Bretagne*, **65**, 97–134

Peacock, D P S, 1967 Forged brick-stamps from Pevensey, *Antiquity*, **47**, 138–40

Poux, J, 1922 *La cité de Carcassonne*, **1**

Roach Smith, C, 1861 Dax (Dep des Landes), *Collectanea Antiqua*, **5**, 226–40, esp 239

RCHM 1962 *An inventory of the historical monuments in the city of York*, **1**, *Eburacum, Roman York*, Royal Commission on Historical Monuments

RCHM 1972 *An inventory of the historical monuments in the city of York*, **2**, *the defences*, Royal Commission on Historical Monuments

Tatton-Brown, T, 1978 Canterbury, *Curr Archaeol*, **6**, 78–82

Thompson, F H, 1973 The gates of Roman Lincoln, *Archaeologia*, **104**, 129–207

von Petrikovits, H, 1971 Fortifications in the north-western Roman Empire from the third to the fifth centuries AD, *J Roman Stud*, **61**, 178–218

Wacher, J S, 1975 *The towns of Roman Britain*

Wightman, E M, 1970 *Roman Trier and the Treveri*

Will, E, 1963 Les enceintes du Bas-Empire à Bavay, *Acta et Dissertiones Archaeologicae*, **3**, 99–110, Zagreb

The reuse of monumental masonry in late Roman defensive walls

T F C Blagg

'Chaque fois que, dans les cours des siècles, on a entamé quelque partie des murailles romaines, des monuments épigraphiques ou figurés sont apparus aux regards étonnés.' (Waltzing 1904–5, ii, 53)

What Waltzing wrote about Arlon might be taken to stand for a general feature of late Roman urban defences in the western provinces. The widespread practice of the builders of these defences of reusing masonry from earlier buildings has, as its unintended consequence, a great enrichment of our knowledge of provincial sculpture and architecture, of religious practices, and of the names and history of the many individuals recorded on funerary and other inscriptions.

In such cases as the funerary monuments from Neumagen (Fig 116; von Massow 1932) the structural remains have been published in full and exemplary fashion. In others, while the epigraphic evidence may have found its way into the *Corpus Inscriptionum Latinarum* or other publications of inscriptions, the architectural material has not received much consideration. This paper cannot attempt a comprehensive study of a very large body of evidence; but some points of interest may be derived, even from a selective survey. This will include sites in Britain, the Gaulish provinces, Germany, and northern Spain. It will also be extended a little beyond the category of purely urban defences, to include military defences in which monumental masonry was reused in a similar way.

It has long been recognized that many of the towns of Gaul were unfortified until the time of the barbarian invasions which began in the middle of the 3rd century. Earlier defences do not incorporate the reused masonry which is among the common characteristics of these latter fortifications (Blanchet 1907, *passim*; Butler 1959, 27). It is easy to form a vision of an urgent response to crisis, in which townsmen hastily seized any materials which lay to hand, either from buildings which the barbarians had laid in ruins, or from those whose functions in a more leisured and peaceful life could be sacrificed to more pressing needs for protection and survival (eg Butler 1959, 44).

Although this generalization does have elements of truth, it conceals such factors as the disparity in date among these fortifications, the nature of the reused masonry in question and the monuments from which it had been taken, and disparate features of the construction of these walls, all of which may suggest other interpretations than what Johnson, in dissenting from such an explanation, has called 'a spontaneous and haphazard reaction to the invasions of the late 3rd century' (Johnson 1973, 223).

The types of monument from which masonry was taken and reused include: tombs and gravestones; temples, religious precincts, and altars; such public buildings as forums, baths, and arches; and milestones. Of these, the funeral monuments were the most usual victims. Sites where other buildings were destroyed or were in such disrepair already as to provide convenient quarries present a less regular picture. There is a significant difference between provinces, for in Spain and Britain it was exceptional for town defences to incorporate material from structures other than tombs, and in Britain even that was not usual. We must, however, examine the material in some detail, because it is not always easy to assign a piece of carving with certainty to a building of particular type.

At Lincoln, for example, inscribed tombstones have been found built into the *colonia* wall at various points (*RIB* 250, 256, 263, and possibly 251, found only ten ft south of the north wall). Plain mouldings, engaged fluted pilasters, and a decorated cornice were found in the bottom courses of the late 4th century west gate of the lower town (Fig 117; Colyer 1975, 239–40 and pls xxxv, xxxviii). Their scale and nature is quite consistent with their having formed part of a funeral monument similar to that of the Secundinii at Igel (Dragendorff & Krüger 1924). Nevertheless, although they are rather too small to have come from a major building, a small temple cannot be ruled out as a possible source. It is tantalizing that the letters inscribed on five blocks recently found near a postern gate in the south wall (Fig 118) do not provide a restorable text which might have indicated the type of building to which they had belonged (Reynolds in Colyer & Jones 1979, 86 and 87, fig 13; *Britannia*, **6** (1975), 284–5). The height of the letters, however (200 mm), is rather greater than one would expect on a funerary monument. The dating evidence for this part of the defences has not yet been published, but in any case the stones formed part of a later refacing of the wall.

London is the only other major Romano-British town where it is certain that monumental stonework was reused extensively in the defences, and that was also in the late 4th century. There are two major groups: the stones incorporated within the external towers at Trinity Place, Bevis Marks, Castle Street, and Camomile Street (RCHM 1928, 99–103 and pls 22 and 30–2); and those which formed the lower courses of part of the riverside wall (Hill *et al* 1980). Most of the material in the first group is funerary in character; the tomb of the procurator Classicianus is probably the best known. The only material which might not be ascribed to such a source is a column shaft decorated with lattice work and overlapping leaves (RCHM 1928, pl 19), which could have formed part of a small votive column, and a number of coping stones which, as Chapman has suggested (in discussion), may have surmounted crenellations

Fig 116 Stones from the funerary monument of Albinius, reused in the walls at Neumagen (photo: von Massow 1932, frontispiece)

Fig 117 Moulding and fluted pilasters built into the north tower of the lower west gate at Lincoln (scale = 20 cm; photo: T F C Blagg)

on parts of the wall demolished when the external towers were built, though Baatz has pointed out (also in discussion) that such stones are also known from temple precinct walls and those of funerary enclosures. The stones from the riverside wall present a totally different picture. In addition to altars and a religious relief, they can be ascribed to two major structures: a monumental arch (Fig 119), and a screen decorated with figures of deities. Here, certainly, buildings of public character appear to have been demolished deliberately to provide material for the wall's construction.

It does not seem altogether clear whether the city wall of Bath in which Camden and Leland saw tombstones and other inscriptions was wholly the medieval structure, or was in part Roman work (Cunliffe 1969, 201–3; *RIB* 161, 164, 166–9). An ashlar wall containing reused Roman architectural material, including two funerary reliefs, was uncovered in 1803, and although it might have been part of the Roman north gate, a medieval date for the structure is not thought impossible (Cunliffe 1969, 202–3 and 207). Another case of possible reuse of stone in a town wall is the milestone found on the site of the north wall of Kenchester (*RIB* 2250).

One other British site deserves particular mention, because in the large-scale use of tombstones in the lower courses of the wall it provides the only real parallel in the province for what we find in Gaulish fortifications. Between 1883 and 1892 nearly 100 funerary reliefs were recovered from the north wall of the legionary fortress at Chester. The only additional items which are not likely to be funerary are a coping stone, a relief of a figure in the attitude of Cautes, and a large pilaster capital (Wright & Richmond 1955, nos 164, 171, 176). Two friezes might have come from a small temple rather than from a tomb (*ibid*, nos 165, 166). The latest datable inscription is of AD 213–22, and heavily weathered, suggesting to Wright and Richmond (1955, 5) that this reconstruction work was not undertaken until the late 3rd century.

With military fortifications of the 3rd and 4th centuries in Britain, as with the towns, the reuse of monumental masonry is relatively uncommon. At Richborough, the west gate and walls of the Saxon Shore fort incorporated dressed greensand blocks, pieces of oolitic limestone, and marble veneers from the enormous Flavian quadrifrons arch which was demolished at about the same time, having already become somewhat dilapidated (Bushe-Fox 1926, 30 and 36; Strong 1968, 42). An altar dedicated by a prefect of the *Classis Britannica* was reused in the approximately contemporary fort at Lympne (*RIB* 66), and in 1804–5 an altar and some milestones were apparently found in the walls of the late 4th century shore fort at Bitterne (*RIB* 97, 2222, 2223, 2226, and 2228). Mr R A H Farrar has also kindly drawn my attention to five stones featuring classical deities in niches, found in robbing part of the wall of the fort at Old Penrith (Haverfield 1913, 194), and other mouldings which appear to have been reused there. These sites contrast with the Romano-British towns and with Chester, in that funerary monuments are not recorded among the reused material, apart from the lion in the north gate at Richborough.

We may now consider some of the evidence from other provinces. Five defensive circuits in Hispania Citerior—at Astorga, León, Saragossa, Lugo, and Barcelona—were studied in a paper by Richmond (1931). The last two, and Gerona, have since been the subject of detailed monographs (Arias Vilas 1972; Balil 1961; Serra Rafols 1942). Apart from tombstones, there is little decorative material to suggest that other buildings were quarried extensively, except at Barcelona. There are five altars among the stones from León, for example, and from Lugo religious dedications, but nothing to indicate a public or private building of any size, as both Richmond (1931, 90) and Arias Vilas (1972, 64) have observed. The situation is somewhat different at Barcelona, where 3rd century modification of the Augustan defences incorporated among other architectural pieces the frieze, architrave, and Corinthian columns of a portico, and where one of the gate towers was built round what has now been identified as the Augustan SE gate (Granados forthcoming; Balil 1961, 82–103). Although excavation of some of the towers in the late 1950s produced the remains of funeral monuments and marble statuary (Balil 1961, 126 ff; Weiss 1961, pls 6 and 7), the incorporation of parts of earlier buildings other than funerary structures suggests that those which stood in the projected path of the defences (Weiss 1961, 192) may have provided a fair proportion of the material reused in their construction. Another such building was that which contained the mosaic floor found beneath part of the defences in the Bajada de Santa Eulalia (Balil 1961, 83–4). The defended area of the town, at 9 ha, is considerably smaller than those of Lugo (34 ha) and Saragossa (60 ha), for example, and comparable with the 6 ha of Gerona, where some architectural material was also reused. In such places, where the circuit was considerably smaller than the built-up area, this kind of reuse is only to be expected.

The same explanation may account for what has been found in some of the Gaulish and German fortifications. At Saintes, some of the column bases of a building adjacent to the walls were left in position, while column shafts and other architectural elements were built into the structure (*Gallia*, **23** (1965), 359; **25** (1967), 251–2). At Bavai, the defences of the *castellum* which was built on the site of the town of Bagacum Nerviorum were planned so as to use the wall of the *cryptoporticus* of the forum, and they contained decorative architectural features which pre-

Fig 118 Inscribed stone moulding built into the south wall at Lincoln (scale = 20 cm; photo: T F C Blagg)

sumably derived from a demolished superstructure (Biévelet 1966). Parts of the defences of Famars and Tournai also made use of the walls of earlier buildings (Mertens, this volume).

In some cases, then, the demolished buildings have been excavated; in others, their nature can be inferred from the character of the reused stonework, and they included some buildings of importance. At Sens, for example, engaged columns and panels carved with vinescrolls and marine scenes were identified by Julliot as having been taken from the Baths (Julliot 1898, 119 ff). The seven or eight bottom courses of the walls of Senlis reused pieces of frieze and cornice and the voussoirs of a decorated arch from a temple of Jupiter (Espérandieu 1955, nos 8352–8357). The foundations of the city walls of Mainz contained carved stones from a Victory monument which may have stood in the legionary praetorium, pilaster capitals and cornices from an octagonal building, the pediment of a small temple of Mars, and the monumental arch erected by Dativius Victor (Behrens 1953/54, 74, 77, 83–84). The *castellum* at Alzey made extensive use of stone from a number of religious monuments. Remains of Jupiter columns were found beneath the east gate of the fortifications, which are datable to the 360s, but Building C, the so-called 'Skulpturenbau', is thought to have anteceded the *castellum*, and included in its lower courses Tuscan columns, reliefs, and altars from a temple of Apollo (Fig 120; Behn 1929/30, 85–96). To take one example from the south-west of Gaul, the walls of Bordeaux included friezes and cornices of a size appropriate to public buildings—'some grandiose monument decorated with marine motifs'—and parts of a water fountain, in addition to a large number of tombstones, altars, and religious statuary (Etienne 1962, 203–12).

Most Gaulish and German towns with fortifications built after the mid 3rd century made use, and in some cases extensive use, of material from public buildings as well as from funerary structures. Blanchet, in his study (1907) of the defensive circuits of Roman towns in Gaul, makes frequent mention of the practice, and Butler summarizes the evidence from some of the sites (Butler 1959, 40–1). It happened not only in towns close to the Rhine frontier, but also in those as far from it as Bordeaux, Périgueux, and Saintes. In this, they differ from most late imperial town defences in Britain and Spain. Can this adequately be explained as a result simply of their being closer to the source of danger (remembering that some of the raids penetrated deep into Gaul), and therefore of their requiring a much more urgent provision of building materials?

Two factors relevant to this question need first to be taken into account: the date, and the quality, of the defensive works. The evidence for dating the walls has been considered by Butler (1959), and more recently by Johnson, in a paper concerned with three of the sites (Johnson 1973), and more generally in his contribution to this volume. In addition to a fairly small amount of literary and epigraphic evidence, the information from stratigraphic excavations is increasing slowly, though there are still many towns where it is either not yet available, or too slight to permit close dating.

There is a substantial number which can be attributed to the last quarter of the 3rd century, notably in Gallia Belgica, where Johnson has suggested that they might represent a concerted scheme begun under the emperor Probus after AD 276 (Johnson 1973, 222). As he has also remarked, the completion of defences might have taken several years, so such finds as a coin of Diocletian in the core of the wall at Beauvais do not necessarily provide a *terminus post quem* for a whole programme of construction, as distinct from the part of the work in which they actually occur. There are, however, defensive works with reused masonry which can be placed firmly in the 4th century. The foundations of the Gautor at Mainz produced coins of Constantine (Behrens 1953/54, 74). Neumagen, '*divi castra inclita Constantini*' as Ausonius called it (*Mosella*, 11), was one of a number of early 4th century *castella* which used material derived from civilian settlements (von Massow 1932). At Strasbourg, refacing of the Trajanic legionary defences comparable to the reconstruction work at Chester was effected under Valentinian I, and included numerous tombstones, the base moulding from a circular building (possibly a temple), as well as a Corinthian capital and cornice fragments (Forrer 1927, 88–9, 151). The reused masonry in the walls of Saverne is likely to have

Fig 119 Perspective reconstruction of the London Arch, from stones built into the Roman riverside wall (drawing: Sheila Gibson)

Fig 120 Architectural stonework and altars incorporated in Building C at Alzey (photo: Behn 1929/30, taf XVII)

been placed there at the time either of their construction *c* 310, or of their repair under Julian in 356/7 (Ammianus xvi, 12). A coin of Constantine II, of 321 or 326, was found mortared into the walls of Auxerre (*Gallia*, **12** (1954), 510). At Lincoln and London, the datable parts of the fortifications which reused monumental masonry also belong to the 4th century.

If, therefore, the employment of monumental masonry from other buildings is to be seen as a response to crisis, it was not a single crisis. Nor, when one considers the second factor, the quality of much of the building work, can it often be seen as a particularly urgent response. This may be deduced from the careful tooling and laying of much of the walling, to which Johnson (1973, 217) has called attention in the cases of Beauvais and Senlis, and from such decorative but functionally unnecessary additions as the mouldings on the interval towers of Barcelona (Balil 1961, 24), the pilasters and cornice mouldings of the tower of the east gate at Périgueux (Blanchet 1907, pl xviii), or the geometrical patterning in stones of contrasting colours in the wall facings at Le Mans (Butler 1958, 34).

A distinction should also be made between systematic and casual reuse of masonry. By systematic, I mean those cases where there is evidence for extensive and purposeful demolition of monuments, and, or alternatively, where the resulting materials have been given a specific place in part of the defensive structure, as in providing the foundation courses of the wall or the core of external towers. Chester, London, the west gates at Lincoln and Richborough, Sens, Bordeaux, Alzey, Arlon, and Neumagen, among the sites already mentioned in this paper, provide examples of this kind of use. By contrast, one may consider as casual the incorporation of material from a building which happened to be in the path chosen for walls in which masonry is not otherwise reused in a regular manner, or sporadic occurrences of mouldings, tombstones, milestones, or other inscriptions, which one may suppose to have been eclectically gathered in assembling building material. Most of the Spanish towns, parts of the walls of Lincoln, Bitterne, and Old Penrith, Beauvais, and Soissons, fall within this category. Clearly, the effect on a community of the loss of some of its amenities was potentially greater where the reuse was systematic. Closer study of this factor is desirable if local variations in circumstances are to be given due weight in testing the validity of generalized explanations.

It has been stated, for example, that 'the reused material found in such quantities when walls of the western type are demolished is evidence that a great and common disaster had affected the towns of Gaul' (Butler 1959, 40). In general terms, and in some particular cases, we are informed of the disaster; thus Aurelius Victor (xxxiii, 3) refers to how '*Francorum gentes direpta Gallia Hispaniam possiderent, vastato ac paene direpto Tarraconensium oppido*'. Even so, we lack literary evidence for the specific nature of the devastation and plundering. Nor is there much archaeological evidence to demonstrate conclusively that the monuments which provided this material had been laid waste by enemy action. Such does appear to be the case at Paris, admittedly, where the amphitheatre, theatre, and public baths show evidence of burning, and where pieces of the same architrave have been found on the site of the forum and built into the walls which fortified the Île de la Cité (Duval 1961, 137, 278–9). It should be appreciated, however, that the demolition of buildings or their damaging beyond repair takes time and energy. Although I would not discount their agency entirely, I find it rather unconvincing to suppose such organization and deliberation of purpose on the part of the barbarian raiders as to account for substantial and irreparable damage on so wide a scale as the reused masonry has been taken to imply.

I have considered elsewhere (Blagg 1981) a comparable source of evidence, that for destruction at Romano-Celtic temple sites in the 3rd century. The number of such sites is not impressive, and at some of them other causes than barbarian attack appear to account for the destruction. Exposed rural temple sites would seem ripe candidates for plunder. Many were deserted in the 3rd century, perhaps in consequence, but few appear to have been destroyed. While, then, it is obvious that the provision of town defences was a result of political and military disturbance, one might consider that the reuse of materials from earlier buildings in those defences followed more from the precautionary measures than from antecedent physical destruction; that is, that the wall-builders themselves were mainly responsible for much of the destruction.

This might have involved making use of material from buildings which were already in disrepair through lapse of time, as was the Richborough arch, as well as cutting losses on buildings which had suffered a degree of damage in the course of barbarian looting. It might have involved the demolition of buildings which had already become redundant. The London arch and the Screen of Gods were probably such monuments, and indeed the treatment of pagan religious monuments as redundant in the 4th century, under Christian emperors, may not be thought too surprising. There is, furthermore, the point already mentioned above, obvious, though so far neglected, that much of the material may have been reused for no more sinister reason than that it came from structures which lay in the path of the new defensive works.

The construction of walls and, equally importantly, wide ditches round existing but previously undefended urban boundaries inevitably involved the clearance of structures sited immediately outside such inhabited areas, a large proportion of which will have been funerary monuments. It is probably no accident that London's external towers included the remains of early funerary structures, such as the tomb of the Flavian procurator Iulius Classicianus, which are likely to have been closest to the urban boundary. Since funeral monuments figure so commonly among the reused stonework, it might be thought that this factor, rather than an arbitrary or sacrilegious disregard of ancestral graves, should be given some weight. Similarly, in those towns where the defended circuit was of smaller extent than the area of the whole, destruction of standing buildings, whether public or private, will have been as inescapable as it has been in the path of urban motorways and inner ring roads in modern towns. The planning of the course of the defences will have involved, in addition to considerations of military tactics, the weighing of priorities concerning which buildings were to be included or excluded, and which to be demolished. We might know more about the nature of these decisions if further research were done on the precise relation of the course taken by the walls to the overall topography of the town, in particular the positions of contemporary buildings and cemeteries.

In making these last few points, I have certainly not intended to replace one monocausal explanation by another. If it were taken as such, it could easily be faulted. The Tour de Vésonne at Périgueux, for example, stood close to, but not in the line of, the walls, yet its marble columns and other decorative features were taken and immured. If the water fountain bequeathed to Bordeaux by the praetor C Iulius Secundus had stood in the way of the city walls, it would be difficult to understand why one fragment of the inscription from it should have been found on the opposite side of the circuit from the others (*CIL*, **13,** 596–600).

I have attempted in this paper to draw attention to some of the variations in the way in which monumental masonry was reused in defensive walls in the western Roman provinces in the 3rd and 4th centuries; to the diversity in the types of structure from which the masonry derived, and in the date and extent of the practice; and to the possible reasons for it, and what has been or might be inferred in consequence about the condition of the towns at the time these defences were built. Some of the suggestions may be fairly hypothetical, but I have felt that the commonly accepted view, that the practice attests widespread destruction of buildings by the barbarians, or an abandonment of them enforced by pressing danger, takes insufficient account of the local and chronological variations which exist, and which must be considered in more detail, in relation to all the features of the construction of these urban defences, if we are to understand them better.

References

Arias Vilas, F, 1972 *Las murallas romanas de Lugo, Studia Arqueologica*, **14**

Balil, A, 1961 *Las murallas romanas de Barcelona, Anejos de Archivo Español de Arqueologia*, **2**

Behn, F, 1929/30 Neue Ausgrabungen im Kastell Alzey, *Mainzer Zeitschrift*, **24/25,** 71–88

Behrens, G, 1953/54 Verschwundene Mainzer Römerbauten, *ibid*, **48/49,** 70–88

Biévelet, H, 1966 L'exploration archéologique à l'est de l'ensemble monumental et au sud de la première forteresse de Bavai, *Revue du Nord*, **48,** 557–77

Blagg, T F C, 1981 Architectural patronage in the western provinces of the Roman empire in the third century, in *The Roman West in the third century* (eds A King & M Henig), Brit Archaeol Rep, **S109,** 167–88

Blanchet, A, 1907 *Les enceintes romaines de la Gaule*

Bushe-Fox, J P, 1926 *First report on the excavation of the Roman fort at Richborough, Kent*, Soc Antiquaries London

Butler, R M, 1958 The Roman walls of Le Mans, *J Roman Stud*, **48,** 33–9

———, 1959 Late Roman town walls in Gaul, *Archaeol J*, **116,** 25–50

CIL *Corpus Inscriptionum Latinarum*

Colyer, C, 1975 Excavations at Lincoln 1970–72: the western defences of the lower town: an interim report, *Antiq J*, **55,** 227–66

Colyer, C, & Jones, M J, 1979 Excavations at Lincoln: second interim report: excavations in the lower town 1972–8, *ibid*, **59,** 50–91

Cunliffe, B W, 1969 *Roman Bath*, Soc Antiquaries London

Dragendorff, H, & Krüger, E, 1924 *Das Grabmal von Igel*

Duval, P M, 1961 *Paris antique*

Espérandieu, E, 1955 *Receuil générale des bas-reliefs, statues et bustes de la Gaule romaine*, **14**

Etienne, R, 1962 *Bordeaux antique*

Forrer, R, 1927 *Strasbourg—Argentorate*, **1,** *das roemische Strassburg—Argentorate*

Granados, J O, forthcoming La primera fortificacion de la Colonia Barcino, in *Papers in Iberian archaeology* (eds T F C Blagg, R F J Jones, & S J Keay), Brit Archaeol Rep

Haverfield, F, 1913 *Voreda*, the Roman fort at Plumpton Wall, *Trans Cumberland Westmorland Antiq Archaeol Soc*, n ser, **13,** 177–98

Hill, C, Millett, M, & Blagg, T F C, 1980 *The Roman riverside wall of London*, London Middlesex Archaeol Soc spec pap, **3**

Johnson, J S, 1973 A group of late Roman city walls in Gallia Belgica, *Britannia*, **4,** 210–23

Julliot, C, 1898 *Inscriptions et monuments du musée gallo-romaine de Sens*

RCHM, 1928 *An inventory of the historical monuments in London*, **3,** *Roman London*, Royal Commission on Historical Monuments

RIB, 1965 Collingwood, R G, & Wright, R P, *The Roman inscriptions of Britain*, **1**

Richmond, I A, 1931 Five town walls in Hispania Citerior, *J Roman Stud*, **21,** 86–100

Serra Rafols, J, 1942 *El recinto antiguo de Gerona, Archivo Español de Arqueologia*

Strong, D E, 1968 The monument, in *Fifth report on the excavations of the Roman fort at Richborough, Kent* (ed B W Cunliffe), Soc Antiquaries London, 40–73

von Massow, W, 1932 *Die Grabmäler von Neumagen*

Waltzing, J-P, 1904–5 *Orolaunum vicus: Arlon à l'epoque romaine; ses monuments et son histoire*

Weiss, A H, 1961 The Roman walls of Barcelona, *Archaeology*, **14,** 188–97

Wright, R P, & Richmond, I A, 1955 *Catalogue of the Roman inscribed and sculptured stones in the Grosvenor Museum, Chester*

As an introduction, I want to draw your attention to the enormous and sophisticated town walls of the ancient east, for instance, the double walls of Nineveh (destroyed 612 BC). These were in existence long before any town wall of similar dimensions and technical quality was known in the west. Furthermore, the Assyrian palace reliefs display many scenes of sieges with attention to detail (Barnett 1959). They are unique sources for techniques of assault and defence and of the tactics of their time, and some of the ageless, elementary weapons and tactics used in sieges over the centuries are represented.

The principles of elementary assault and defensive weapons and tactics

In sieges, long-range weapons were of course of the greatest importance. As reliefs and sometimes vase paintings show, and written sources confirm, the defenders used javelins, bows, missiles hurled by slings, or stones thrown by hand. Which of these was preferred in a siege was a matter of the military traditions of the period.

Attackers employed bows, slings, or javelins, the last being of use against comparatively low defences only. They would try to climb the walls on scaling ladders, or break down the gates or even the walls using battering-rams. Undermining walls or demolishing them were also old techniques. To achieve these ends the attackers had to approach the wall under the fire of the defenders. A simple counter tactic for a group of attackers was to build a *testudo* by closing their shields over their heads (Fig 121).[1] Another tactic was to split up into two groups, one holding the defenders down by employing long-range weapons. Especially useful were bows or slings; in later and technically more advanced warfare, this was the usual task of small torsion arrow-shooting artillery. Under cover of the fire, the other group could then move against the wall equipped with scaling-ladders or a battering-ram (Cichorius 1896, pl 24).

A very old and important—but in modern literature much neglected—defensive weapon was the stone thrown by hand. The stone was of course an extremely economic weapon and quite effective, especially if thrown from a high wall. Thus the throwing of stones from town walls by defenders was common practice over the millennia up to the late Middle Ages. Different sizes of stones were used, and there were two methods of throwing them. Bigger stones were dropped using both hands to hit a target near the foot of the wall; smaller stones were hurled using one hand over a range comparable to that of a javelin (Fig 121). Both techniques are shown on the Assyrian reliefs, were known to the Greeks (Fig 121),[2] and are to be found depicted on Roman historical reliefs, for instance Trajan's Column (Cichorius 1896, pl 84, 99). Late medieval illustrations of sieges sometimes show walls still being defended by stones thrown by hand even against early artillery pieces. There are also many literary references including Roman military sources (Vegetius, I, 16: IV, 8). This particular defensive technique required heaps of worked stones of similar shape and weight,[3] with the weight regulated according to the method of throwing. A store of stones of the correct weights was a necessity, because during an attack a high rate of accurate fire must often have been of vital importance for the survival of a town. This cannot be achieved with unworked stones of uncertain weight and there are many archaeological finds of more or less well worked stone balls from ancient town and fort defences. Often they are interpreted as *ballista* shot, but in reality nearly all of them will have been worked stones for throwing (Baatz 1966, 202–3; 1973, 112, pl 28, 1; Erdmann 1977, 80–2, pl 16).

Crenellations

There are not many crenellations left on ancient walls. An example of a Greek crenellation is preserved on a tower of 4th century BC date at Messene, published recently by Haselberger (1979, 101, fig 5, 109, fig 7). Other crenellations of classical or Hellenistic date are known from Lilaea, Phyle, Caunus, and Dura (Lawrence 1979, 357–62). Their number is insufficient to draw any general conclusion. The width of the embrasures varies from 0·60 to 1·45 m.

The differences in embrasure width reflect the use of different weapons. It would have been difficult to thrust a javelin, hurl a missile from a sling, or throw a stone by hand through a narrow embrasure: a defender would have risked hitting his hand against a merlon. However, a wall with narrow embrasures gave better cover against enemy fire. Consequently, narrow embrasures were employed when suitable weapons such as bows or (in later periods) arrow-shooting catapults were used.

Another type of *epalxis* on Greek walls was the closed battlement, which sometimes carried a roof. Closed battlements were provided with rectangular windows or sometimes narrow slits for the use of defensive weapons (Lawrence 1979, 367–70). No closed battlements have been found on Roman defences so far.

On Roman walls there are likewise only very few crenellations left. An important example is preserved at the Praetorian Camp in Rome. The first Tiberian wall was heightened at successive periods and part of it was later built into the Aurelianic defences (Richmond 1927). The early Tiberian embrasures are especially wide (embrasures 2·9 m, merlons 0·6 m). Obviously this relates to the contemporary weapons used by the Roman army. *Sagittarii* were not common and were only found in a few specialized auxiliary units, therefore camps and forts of the Roman

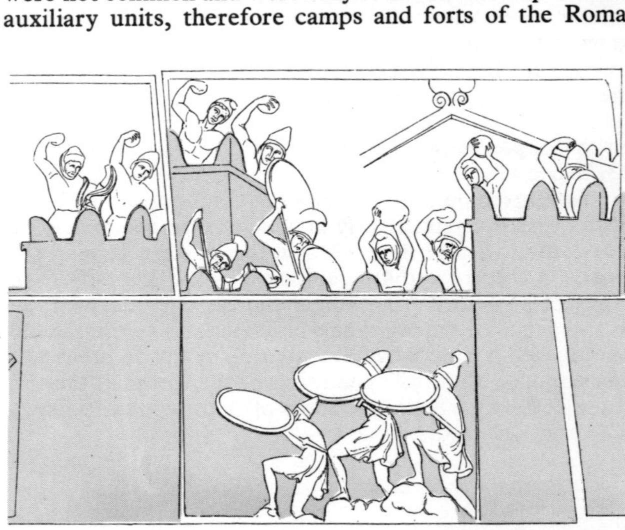

Fig 121 Heroon of Trysa (first half of the 4th century BC), siege of a town (after O Benndorf)

Fig 122 Saalburg, Bad Homburg, Germany: reconstruction of merlon with traverse and coping stone after finds from the upper German limes

army would have had wide embrasures for the employment of javelins and hand-thrown stones (Cichorius 1896, pl 24; Richmond 1935, 27–9, fig 10). Comparatively wide embrasures are also known from early Roman town defences, for instance at Fondi (2·4 m) and Pompeii (1·5 m). In later battlements the embrasures tended to become narrower (Verona, about AD 265, 1·2 m). In late antiquity they were often even narrower and this points to the extensive use of the bow, the *manuballista*, and other small catapults.

Sometimes merlons were provided with traverses (Fig 122) which gave more stability to the battlements, especially against stone-shot from *ballistae*. However, this device antedates the invention of ancient artillery (Lawrence 1979, 361). Therefore, the threat of artillery could not have been the only reason for building traverses. More general tactical reasons plus slightly improved cover for the defenders must have led to their introduction (Lawrence 1979, 360–1).

Galleried curtain walls

To be a serious obstacle against scaling-ladders, a town wall needed to be at least 7 m high at the level of the wall-walk, which meant about 9 m to the top of the merlons. To overcome a wall of this height, attackers would have to construct siege-ramps or movable towers, to try to move big, protected rams against the defences. All these devices and machines were well known in the ancient east and are represented for instance on Assyrian reliefs. In the Hellenistic period methods of attack were developed to perfection, technically. The invention of torsion artillery during the 4th century BC (Marsden 1969, 48–64) brought an additional advance to siege techniques.

Arrow-shooting catapults and archers brought into action from the upper storeys of siege-towers could easily dislodge defenders from the wall-walk. One reaction of military architects was to construct closed battlements with roofs, mentioned above. Another response was the provision of galleried curtains, which may go back to early eastern military architecture (Naumann 1971, 238–9, figs 305, 306; 309–10, figs 412–15). The galleries were provided with slits for firing from and the defenders were protected almost completely, but only special weapons such as bows and small torsion engines could be employed. Two Hellenistic galleried walls are still standing at Side and Perge (Lawrence 1979, 370–5); the device is recommended by the contemporary technical literature also (Philon, *Parasceuastica*, 80–1). Galleried curtains were adopted by the Romans, and the Aurelianic walls of Rome are an outstanding example (Fig 123; Todd 1978, 29–30, 48–50). They were still in use in the late Roman and Byzantine periods, and the land wall of Constantinople is a most impressive structure having a galleried wall in the second line of defence (Meyer-Plath & Schneider 1943, 84–92, fig 21).

Ditches and forewalls

In ancient warfare, fighting in tightly closed lines was normal, and this also applied if attacking a town wall. Soldiers had learned to cover each other with their shields, building a *testudo* if necessary. Therefore, the main tactical purpose of a ditch in front of a wall was to break the closed line of the attackers and their cover of shields at a point where the weapons of the defenders had maximum effect. The distance of the outer edge of the ditch (or ditches) from the wall, consequently, was dependent on the type of arms used by the defenders. If they employed javelins or stones thrown by hand, a short distance was chosen, rarely more than 15 m. Archers or slingers were effective over a longer range, therefore defences regularly manned by them had ditches lying farther out. This was also the case with walls well supplied with torsion engines.

A ditch also had the effect of heightening a wall as a measure against scaling-ladders and, especially if filled with water, it impeded or prevented the undermining of a wall.

Wide and deep ditches were a substantial obstacle to machines of all sorts and siege-ramps. They would have to be filled in by the attackers first, and to hinder this such ditches were sometimes defended by a forewall (*proteichisma*). The forewall itself would act as a further obstacle to machines. Wide and deep ditches combined with a forewall were designed especially against highly organized attackers, who could construct siege-ramps and bring big machines into action against the defences. At an early date they were used in defences of the ancient east (for instance, Nineveh). Later they were adopted by the Greeks, and some of the more important Hellenistic towns were provided with *proteichismata*. No example of this strong and refined type of defences is definitely known so far from the west. In the east, a number of the old Hellenistic town walls continued in use over many centuries, and new defences provided with *proteichismata* were constructed in the late Roman and Byzantine periods. The most impressive and best preserved example still standing is the land wall of Constantinople.

Towers

Towers were built for many purposes: they allowed more defenders to man the defensive circuit, they were vantage points, they provided room and sometimes accommodation for patrols and guards, and they could be used as dry store-rooms for arms, war-machines, and provisions. Even if the towers did not project in front of the wall, they offered

Fig 123 Reconstruction of the second, Maxentian phase of Aurelian's wall at Rome

all the functions mentioned above. Therefore, a tower which did not project forward was by no means useless.

If the towers did project forward, flanking fire along the curtain wall was possible. This must have been very effective if a mass attack was made because the attackers could not be fully protected by their shields if under fire from two or more directions. A basic requirement was, of course, the employment of long-range weapons from the towers (bows, slings, or catapults). Only if a town could provide enough skilled men trained in such arms would projecting towers be of any use in this respect.

There was a lack of archers and slingers in most units of the Roman army during the 1st and 2nd centuries AD. This was presumably one of the reasons why most Roman camps and forts of this period were not provided with towers which projected enough to allow flanking fire. Very few Roman towns at this time had military personnel trained in the use of long-range weapons.[4] This must have been the main reason for the construction of town walls during this period without provisions being made for such flanking fire.

Early town walls in the west

To provide an outline of the architecture of town walls during the early period in the west, some typical examples may be sufficient. An interesting early town wall was built at Cologne in Germania Inferior during the reign of Claudius (see Hellenkemper, this volume). The wall was 2·30–2·50 m thick and nearly 8 m high from ground level to the wall-walk, high enough to be a real barrier against scaling-ladders. The towers were circular, about 9 m in diameter, and projected symmetrically on both sides of the wall. Therefore, the forward projection of the towers was limited and flanking fire must have been nearly impossible. In spite of this, the defences of Colonia Claudia Ara Agrippinensium were a substantial obstacle against barbarians without machines and without the capability to organize siege works. Only a few years after completion the defences proved their value during the difficult episode of the Batavian revolt. Their mere existence may have saved the town from being sacked (Tacitus, *Hist*, 4. 63–5).

The town wall of Aventicum (Avenches) in Germania Superior is thought to be Flavian (Schwarz 1964, 35–

6) and was 2·4 m wide. According to findings at the Tornallaz—where the only surviving tower stands—a length of wall was reconstructed there with a wall-walk 5·60 m above the ground; the ancient wall may have been even lower (*ibid*, 17, n 31). The modest height of the wall was a limiting factor on its defensive value. A single ditch (3·8 m wide and 1·6 m deep) was discovered 2 m in front of the wall and was an element of this defensive system. In the ditch many coping stones from the crenellations were found, and allowed the reconstruction of merlons 1·1 m wide (with tranverses similar to Fig 122) and embrasures 1·9 m wide. As already mentioned, this type of crenellation with wide embrasures was designed for defence mainly by javelins and stones thrown by hand, to which the distance of 5·8 m from the wall foot to the outer edge of the ditch was quite suitable. There were also semicircular towers with a diameter of 7 m, which were set against the inner face of the wall and did not project forward at all. Flanking fire along the curtain was therefore impossible.

At Augst (Colonia Augusta Raurica in Germania Superior) only two lengths of wall were built, flanking two of the main entrances to the town. Nevertheless, semicircular towers were set against the internal faces as at Avenches. These defences are possibly of 2nd century date but were never completed. The walls may have been erected for show only, to indicate the limits of the town. They were clearly not constructed for serious defence, but were a symbol and suggest that there was actually very little danger at the time they were constructed (Laur-Belart 1966, 29–32).

The town wall of Trier (Colonia Augusta Treverorum in Gallia Belgica) dated to the end of the 2nd century was a more substantial structure (Lehner 1896; Koethe 1936; Cüppers 1973). The foundations, up to 3·5 m deep, carried a carefully built wall 3 m wide. At the Porta Nigra, one of the best preserved Roman town gates, the height of the wall-walk above ground level was 6·13 m. Two ditches were found in front of the wall, and a third was found fronting the south wall only. The date of the ditches, especially the outlying ones, is difficult to establish. Possibly the second and third ditches were later additions. The circular towers of 8·7–10·5 m diameter were set symmetrically on the wall, similar to the towers of Cologne which may have acted as a model. Consequently, the possibility of flanking fire along the curtain must have been poor. In spite of this, the defences stood the test during the siege of AD 197 and with the help of the 22nd legion from Mainz the citizens were able to defend their town (*CIL*, **13,** 6800).

Later town walls in the west

The Germanic incursions during the 3rd century AD seriously affected the western provinces. Towns without defences were sacked, ineffective defences were stormed, and as a result some towns were even wiped out completely. This was the case for instance at Augst and Avenches, where, respectively, defences had not been properly built and the existing ones proved too weak. The few towns with efficient defences, however, for instance Cologne and Trier, had to put up with losses also, but life went on and they have survived as important cities up to the present day.

In the second half of the 3rd century, after the central government had been able to stabilize the empire once more, a gigantic building programme was put into action. Dozens of towns near the *limes* and inside the frontier were fortified, many of them according to a uniform scheme which points to central planning. The details of this planning, its financing, the contribution of the state on one side and of the citizens of a town and the *civitas* as a whole on the other, are to a large extent unknown and should be a focus of further research.

Even Rome, the capital of the Empire, was fortified after having been an open city for centuries. The Aurelianic walls guaranteed the survival of the city and its importance as a metropolis beyond the Roman period. The details of construction and the different building phases have been dealt with elsewhere (Todd 1978, 29–30, 48–50; this volume), but it should be emphasized that the walls of Rome, especially after the second, Maxentian, phase, were the strongest and most sophisticated defences in the west. They were the only ones with galleried walls (Fig 123), a device designed to counter organized siege-works and, technically, highly skilled attacks, as stated above. The barbarians were not capable of such attacks, so this particular detail of the Aurelian wall must have been directed against the skill of the Roman army itself. This may have been, as a consequence of the crisis in the 3rd century, a measure against rebellious generals marching against Rome from their provinces. But even the Aurelianic wall does not display the whole range of defensive architecture known in the east; there is for instance no substantial ditch defended by a *proteichisma*.

The new 3rd century town walls in the western provinces often enclosed only a fraction of the original area of the town (Grenier 1931, 403–29). This may have been due to a declining population, but the walls also gave a considerably shorter circuit to defend with the possibility of putting more men on to a certain length of wall than on the older, much longer circuits. Where preserved, the curtain walls are very high, the crenellations and wall-walk being out of range of scaling-ladders. The walls also were quite thick and must have been impossible to breach by the limited means at the disposal of the invaders. Big towers were built which projected effectively, facilitating a strong flanking fire. In addition more towers were now erected. The intervals between towers had been about 100 m (Cologne), 70–90 m (Avenches), and about 90 m (Trier), but by this period approximately half the distance had become standard. A high rate of fire, more accurate over the shorter distance, was the consequence. Multiple or single, wide, outlying ditches point to a shift in defensive weapons in favour of long-range arms such as bows, small torsion catapults, or early types of crossbows.[5] The defences, or at least the towers, must have been manned by professionals trained in the use of these weapons. It is known that there were military units in some of the late Roman towns, and the high percentage of *sagittarii* in the army at this time and the specialized *ballistarii* units also must have been organized with regard to the defence of the towns. But for many towns, the organization of the manning of the defences is still a question needing to be researched.

Big, stone-throwing, torsion artillery would have been of little use for the defence of the towns in the west in the late Roman period. Defenders employed heavy artillery in the first line against the machines of the enemy and against siege construction works. No such machines or works were used by the Germanic invaders of the 3rd and 4th centuries AD. Against mass attacks heavy artillery had only limited effect because of the quite low rate of fire. Therefore in the west only a few army units or towns—if any—would have been equipped with such specialized and expensive weapons.

The change in defensive architecture must have been in answer to the assault tactics of the invaders. Lacking the necessary organizing abilities and the technical means for a regular siege, they would have tried to force the walls by mass attacks using scaling-ladders or simple rams to break the gates. The new types of defences were obviously designed to counter this method of assault. They proved to be quite successful. If defended energetically, they were nearly impregnable. In many cases, they lasted more than a millennium, enabling urban life to continue long after the Roman Empire had fallen.

Conclusion

Developments in the town (and military) defences of the west were related to the growing danger of Germanic invasions: the defences were always adapted to the tactics and weapons of the aggressors. Consequently, many of the architectural devices used against machines and siege-ramps, well known in the east for some time, are not found in the west, with the exception of Rome. There was no technical progress in a modern sense, no typological 'Darwinistic' development, but intelligent adaptation to the requirements of the time of possibilities already known for centuries.

Notes

1 Also on Trajan's Column (Cichorius 1896, pl 51) and on the column of Marcus Aurelius (Petersen *et al* 1896, pl 62).

2 The siege scenes on the monument of the Nereids at Xanthos (about 400 BC) also show stones being thrown.

3 A heap of worked stones is shown behind the parapet of a town wall on the lower frieze of the Francois vase (Arias & Hirmer 1962, 291, pl 44).

4 The Roman state was often very reluctant to tolerate seemingly harmless associations of citizens, for instance a fire brigade (*collegium fabrorum*) for fear of subversive activities (Pliny, *Ep*, x, 33–4).

5 Small hand held torsion weapons of *cheiroballistra* type or the somewhat bigger catapults of Orsova type must have been very common (Baatz 1978, 9–16). Furthermore, a developed version of the old *gastraphetes* may have been in use as a sort of crossbow.

References

Arias, P E, & Hirmer, M, 1962 *A history of Greek vase painting*

Baatz, D, 1966 Zur Geschützbewaffung römischer Auxiliartruppen in den frühen und mittleren Kaiserzeit, *Bonner Jahrb*, **166**, 194–207

——, 1973 Kastell Hesselbach, *Limesforch*, **12**

——, 1978 Recent finds of ancient artillery, *Britannia*, **9**, 1–18

Barnett, R D, 1959 *Assyrian palace reliefs*, Prague

Cichorius, C, 1896 *Die Reliefs der Traianssäule*, Berlin

Cüppers, H, 1973 Die Stadtmauer des römischen Trier und das Gräberfeld an der Porta Nigra, *Trierer Zeitschr*, **36**, 133–222

Erdmann, E, 1977 *Nordosttor und persische Belagerungsrampe in Alt-Paphos, Waffen und Kleinfunde, Ausgrabungen in Alt-Paphos* (ed F G Maier), **1,1**, Konstanz

Grenier, A, 1931 *Manuel d'archéologie Gallo-Romaine*, **1**, Paris

Haselberger, L, 1979 Dächer griechischer Wehrtürme, *Mitteilungen des Deutschen Archäol Instituts Athen*, **94**, 93–115

Koethe, H, 1936 Die Stadtmauer des römischen Trier, *Trierer Zeitschr*, **11**, 46–74

Laur-Belart, R, 1966 *Führer durch Augusta Raurica*, 4 edn, Basel

Lawrence, A W, 1979 *Greek aims in fortification*

Lehner, H, 1896 Die römische Stadtbefestigung von Trier, *Westdeutsche Zeitschr*, **15**, 211–40

Marsden, E W, 1969 *Greek and Roman artillery: historical development*

Meyer-Plath, B, & Schneider, A M, 1943 *Die Landmauer von Konstantinopel*, pt 2, Berlin

Naumann, R, 1971 *Architektur Kleinasiens*, Tübingen

Petersen, E, von Domaszewski, A, & Calderini, G, 1896 *Die Marcus-Säule auf der Piazza Colonna in Rom*, München

Richmond, I A, 1927 The relation of the Praetorian camp to Aurelian's wall of Rome, *Pap Brit Sch Rome*, **10**, 12–22

——, 1935 Trajan's army on Trajan's Column, *ibid*, **13**, 1–40

Schwarz, G Th, 1964 *Die Kaiserstadt Aventicum*, Bern

Todd, M, 1978 *The walls of Rome*

PART IV CONCLUSION

Some thoughts on Roman urban defences in the west

J S Wacher

In those happy, euphoric days two or more decades ago, most archaeologists who were interested in the subject thought that they knew a fair amount about urban fortifications in Britain. Admittedly, fashionable theories were propounded which, for a time, held sway, only to be in turn replaced by other equally plausible explanations. Most of the hypotheses which were related to the construction of these defences then came neatly wrapped in convenient parcels and it was not too difficult to fit comfortably the evidence then available into suitable historical contexts. Scholars who were then leading the investigations might now be forgiven if they tend to feel and act not unlike the Ancient Mariner, and it is not perhaps entirely surprising when one of them fixes with his glittering eye some more recent recruit to this field of research who comes springing fully armed, supposedly with all the necessary evidence, to destroy cherished and established theories.

Certainly the last twenty years have seen a rapid expansion in the amount of detailed knowledge available on both sides of the Channel; more sites have been investigated and much more work has been carried out even on those towns where it was once thought that sufficient information had already been obtained. Certainly, also, as the result of this massive accumulation of evidence, it is no longer always possible to support or to propose the simplistic hypotheses that were current not so long ago. But it must be said with emphasis that this welter of detail could be obscuring the main issues. It is now very much the trend, as pleaded by Butler (p 125), to treat each town or minor defended site as an entity, unrelated to its neighbours and with its own distinct and separate chronology for its fortification. Such a tendency appears to be almost as dangerous as some of those fashions which were earlier in vogue, for there still remain certain fixed points about which all researchers must revolve, and there seems no better place to restate them, along with a consideration of both new discoveries and past hypotheses, than at a conference devoted to the subject, so as to give a sense of perspective to its published proceedings.

The papers presented to the conference ranged over a remarkably rich and diverse series of topics, both in time and place. Basically, they fell into three main categories, or combinations of them: historical, structural, and functional. They covered urban fortifications in the provinces of Britain, Gaul, Spain, Germany, and Pannonia; they drew on material from Rome and from the eastern provinces; they looked at systems of both early and late defences, at the methods of manning them, and at the tactics and strategy employed. Also of interest were the contributions on individual sites, such as Xanten, Cologne, London, and Silchester, each with its own special problems.

One thing emerged from these deliberations with great clarity to reinforce what many archaeologists have been arguing for years: the remarkable insularity of Britain in the matter of urban defences, which seldom find comparisons, at least in other western provinces. Indeed no western province has yet produced the complexities which have been observed at some British towns, and it must now be less likely that this is the result of too-limited investigations on continental sites, as might once have been claimed. No other province can appear to match the widespread distribution of urban earthworks in Britain; indeed Cologne, as mentioned by Hellenkemper (p 20), appears as the only example of a town across the Channel to be fortified in such a way, but the circuit is neither well known nor accurately dated. Another, even less well known example may be Rheims (Hassall, p 1). Moreover several British towns such as Verulamium (Frere 1964a; b) and Silchester (Fulford, p 85) possessed more than one circuit of earthworks, each belonging to a different period, while the three *coloniae* of Colchester, Lincoln, and Gloucester, as shown by Jones (p 90), exhibit complexities of a different nature, influenced possibly by the defensive circuits of the earlier fortresses. In this context we should not forget Cirencester which, although not a *colonia*, shows similar complications in the development of its fortifications (Cullen 1970; Wacher 1975, 301–3). Interesting, therefore, is the information now provided from Xanten, but only after extensive excavations have taken place (Precht, p 29), which discloses a similar lack of homogeneity in the curtain wall. Indeed the description of these irregularities helped to stress yet again the important point that excavators should now no longer expect to find homogeneous urban fortifications and that they should be prepared to reveal departures from the norm from the very beginning. Nor can we any longer accept results from one quarter as being representative of the whole circuit.

But perhaps the most crucial legacy which the conference has left us is that it revealed large areas where our knowledge is sadly deficient or even non-existent. Seemingly, therefore, the most useful part which these conclusions can play is to seek to identify some of these areas for the benefit of present and future students working in the field of urban fortifications. This represents, as it were the 'negative' part of the conference, matters not specifically referred to in its many papers.

A good point from which to start is perhaps that covering the legal requirements enforced by the two entries in the *Digest* which relate to urban fortifications (I, 8, 9 (Ulpian); L, 10, 6 (Modestinus)). Both date to the principate of Marcus Aurelius and prohibit construction of defences without the prior permission of both the provincial governor and the emperor. It would be interesting to know what lay behind these legal enactments. Were they enshrining in the law what had hitherto only been accepted custom and, if so, what had made it necessary to legislate in this formal manner? Were they, on the other hand, enacted because provincial governors, or even individual cities and towns, had been taking the initiative in erecting defences and, consequently, in manning them, had breached the *lex Julia* with its prohibition on the carriage of arms? Presumably once enacted the new laws would have been in force throughout the empire, even though they may have been introduced in much more specific circumstances and perhaps relating mainly to the Danubian wars of Marcus Aurelius. These are questions which it is almost impossible to answer but which are vital for the full understanding of urban defences. Moreover, we cannot be certain what the legal position was before the provisions became law; this leaves the defensive circuits of the 1st and earlier 2nd centuries in a legally indefinable position.

Most early circuits were confined to *coloniae* or *muni-*

cipia such as Xanten (Precht, p 29), Lincoln, Gloucester, and Colchester (Jones, p 90), Cologne (Hellenkemper, p 20) and the Augustan cities of Narbonensis, and those in Spain and north Italy. But the early earthworks in Britain at Silchester, Verulamium, and Winchester should not be overlooked (Wacher 1978, 96). It seems inconceivable that some form of imperial permission was not required for their construction, even if the three British examples just quoted lay within the client kingdom of Cogidubnus.[1] But we are probably on safer ground when considering the later circuits; they must surely have been constructed within the framework of the law, and all excavators would do well to remember its prohibitions. It can probably be agreed that the monumental scale of urban fortifications will almost certainly have required, at some time or other, local structural alterations, modifications, rebuildings, or repairs, ultimately giving rise to the complex sequences which are now to be seen round many towns. But somewhere in the labyrinth of structural details there must lie the core of the initial, planned circuits which, if dating from the late 2nd century onwards, should have been built in accordance with the law and so probably required a significant reason for their construction, a reason sufficient to convince the emperor of their need. Consequently it may still be right to attempt to attribute such primary foundations to a specific historical context which can usually be related to some threat of war or rebellion, and, for this reason, it behoves every excavator to seek out and positively identify the primary work and distinguish it clearly from secondary or tertiary modifications. In these circumstances also it is probably still equally right to associate the date of construction of defences of one town with others in its vicinity or in the same province, as for all there was most likely to be one common cause. Consequently, even though we need to know about each alteration, modification, or addition that was made to defensive circuits, it is still vitally important that, ultimately, each should be related to its historical context, otherwise the fortifications will have little meaning. Excavators should not need reminding of this need to equate structural works to the historical record. In this respect it is perhaps worth repeating the old maxim about the dating of defences, since there seems to be a growing tendency now to ignore it: in the absence of epigraphic evidence giving the date of construction, it can only be assessed from the date of the associated artefacts, even though these will never provide more than a *terminus post quem.* With sufficient excavation a time may come when this *terminus* may well approach the actual date of construction, although the excavator can never know this, and there must always remain the possibility that further investigation may advance its date. It is probably also much safer now to consider only material securely sealed beneath a rampart, or in the foundations of the wall, for the *terminus post quem* of the initial construction; finds recovered from the material of the bank or from the wall superstructure may have come from an unidentified repair or modification.

Another area where our knowledge is minimal is that which covers the whole question of cost and financial provision, and which was touched on, but not elaborated, by Webster (p 119), Johnson (p 69), and Casey (p 123). It is extremely difficult to estimate the cost of the erection of massive defensive walls with all their subsidiary works; the key question is whether the cost was met by the community, or wholly or in part from government funds. We do not know, but it is possible to envisage, the economic damage which would have been inflicted on a city or town if they had to meet the full cost themselves. We might start using logistical methods: calculations could probably be made with no great difficulty on the quantities of stone, sand, lime, and gravel which had to be quarried and treated. The identification of quarry sites which, in view of the volumes extracted, would not be insignificant, should indicate the distances over which the material had to be carried and therefore provide some clues to the transport costs. Estimates, backed perhaps by experimental reconstructions, such as those carried out at Vindolanda (Birley 1977, 158–65), The Lunt (Hobley 1967; 1975), or Xanten (Precht, p 29) could be made for the number of man-hours needed for the work (see also Hanson 1978; Wacher 1978, 174, 181). In the end we might achieve some interesting and probably highly revealing figures.

Another area of interest which has largely lacked study as an entity is that which covers unfinished schemes of fortification or those which were modified radically during construction; the number of known examples is increasing. In Britain there is the well known Fosse earthwork at Verulamium (see above) now dated by Frere to the mid 2nd century.

We might wonder also, in view of the report by Fulford on Silchester (p 85), whether the so-called Outer Earthwork there falls into the same category. Examples where masonry was allied with earthworks are seemingly becoming more numerous in Britain and suggest a radical change in design during construction. This association at Cirencester has long been known, as well as that at Exeter, while there are strong reasons for supposing that the London and Chester Gates at Verulamium are contemporary with the unfinished Fosse (Wacher 1978, 97). Fulford has recently shown (p 86) that the main gates at Silchester are related to the latest earthwork, an association which has long been suspected. Less well known perhaps is Dorchester, Dorset, where excavations produced evidence, on the southern part of the circuit, for a wall foundation, 3 m wide, inside the line of the later curtain wall; the foundation was never used but was covered by the tail of the later rampart (Wilson 1973, 315). In other provinces, recent excavations at Xanten (Precht, p 29) have revealed sections of curtain wall of different gauges, 1·75 m and 3·50 m wide respectively at foundation level; it is suggested that there was a break in construction, although the date of the narrow wall has not yet been determined. Johnson reminded us (p 69) of the incomplete masonry walls at Augst in Gallia Belgica and at Lienz in Noricum. There must have been sound reasons to explain the unfinished or modified state of these works, but we are still a long way from identifying them; indeed more examples may yet come to light.

Turning to the later period, no satisfactory explanation has yet been advanced to explain the extremely spasmodic, and very late, provision of external towers for British towns, and the contrast that there is in this aspect of fortification with Gaul and Germany and most other provinces (Hassall, p 1). Can we again detect in Britain an insularity as well as an unfinished system of fortifications? The long accepted view, that these towers were constructed for the more efficient deployment of artillery (Richmond 1930; Corder 1955, 35) provided something of a solution, since, presumably, such towers would only have been constructed where and when artillery was available, and the supply of these weapons was probably very limited. But now that Baatz has made his interesting proposition (p 136) that it is more likely that towers were intended to provide forward concentrations of missile-throwing defenders, their uneven distribution in Britain is less easily explained. For instance, the two towers at Verulamium can now hardly be seen as anything other than an unfinished scheme.[2] What happened to cause its abandonment?

Aldborough

Equally, it is extremely difficult to explain why Caistor-by-Norwich should be the only British town to display two different forms of external tower, although such mixtures are not unknown on the continent and are most remarkably demonstrated at Salona.[3] How also do we explain those sites which were never given towers, such as Silchester or Catterick?[4] As far as Britain is concerned, it may be important to note also the distribution of different forms of external tower: multangular examples tend to be found only in the south-west;[5] semicircular, or variants thereof, tend to be concentrated in the south-east;[6] while rectangular examples occur in the east and as far north as Aldborough.[7] There are exceptions, such as the remarkable fan-shaped towers of Ancaster (Todd 1981, 29, 40), reminiscent of those in eastern parts of the empire (such as the Constantinian fort at Drobeta) but in the main the distributions appear to be worth study, and add support to Johnson's suggestion (p 72) that the defences of certain groups of towns may have been constructed by the same builders.

Apart from the major problems considered above, there are also numerous lesser areas where research could be carried out with profit. We might consider, for instance, the association of town walls and sewers, which in many cases penetrated them to seek outlets into ditches or rivers. Some, such as that which emerges through the north-east gate at Colchester, were sufficiently large to admit a small man.[8] Were these protected in any way, for surely they represented weak points in the circuits? Little is known of water-gates; one must have existed where the Walbrook emerges from the centre of London to join the Thames. Was it protected by flanking towers from which perhaps a heavy chain could be hung across the stream, as was probably used across the canal at Fréjus (Grenier 1931, 301)?

There are also numerous examples in other provinces, as well as in Rome itself,[9] where aqueducts impinge upon, were incorporated in, or made use of, fortifications: Fréjus (Grenier 1960, 47), Nîmes (*ibid*, 96), Lyons (de Montauzan 1908), to name but a few (Blanchet 1908, *passim*); they certainly deserve study. Little is yet known of similar associations in Britain apart from the well attested *castellum* on the north wall at Lincoln.[10] There is a strong suggestion that a *castellum* also existed in conjunction with the Verulamium Gate at Cirencester (Wacher 1961, 66, fig 1), while preliminary indications would suggest their existence in several other towns, including perhaps Leicester,[11] Dorchester,[12] and Verulamium.[13] The point at which the aqueduct at Wroxeter crossed the line of the defences has also still to be examined (Webster & Hollingworth 1959, 133). Most other major towns in Britain possessed aqueducts, all of which at some point crossed the fortifications, and at none of them is it known how this was achieved. The problem tends to be different in Britain where the water supply arrives at most town boundaries either at or below street level, whereas many of the continental aqueducts carried elevated supplies which could be accommodated without making breaches in the walls at low level. On other individual structures, much more knowledge might be obtained by a systematic study of gates, of which a remarkably wide variety of plans are known both in Britain and in other provinces (Wacher 1975, 71–7; von Petrikovits 1971, 200), of internal towers, which have been less well examined than their external counterparts, or of minor features like *ascensus*, of which the two examples at Cirencester can hardly be the only representatives.[14]

But wherever we look to begin our investigations, fundamentally we must always come back, as most people would now agree, to the first and most important need, which is still to find out all that there is to be known about every individual circuit: its walls, banks, ditches, gates, and towers. But it is of paramount importance to remember that that in itself is not enough; it is only the beginning of the complete investigation. Urban fortifications are, by their very nature, a major part of the historical scene; no other structure can probably be so closely related to the march of events. Consequently, without the synthesis and association with the course of history that must follow, the investigations are likely to prove entirely barren and we might as well cease to excavate such structures!

Notes

1 It would have made little difference whether or no Cogidubnus was an imperial legate (Bogaers 1979). As a client king he may have had freedom to act in this capacity; alternatively, if imperial permission was required, it would not have been difficult for him to convince Vespasian of the seriousness of the plight of the British province.

2 Most accounts of the defences at Verulamium have assumed a complete circuit of external towers. It is difficult to see how they have not been found if they once existed (Wheeler & Wheeler 1936, 59).

3 Most conveniently seen for Salona in Wilkes 1969, 360.

4 Lengthy investigations have revealed none at Silchester and very clear aerial photographs of Catterick show no evidence for them.

5 eg Caerwent (Nash-Williams 1930, 263–8) and Cirencester (Wacher 1961, 68; Brown & McWhirr 1966, 244; Cullen 1970, 236). In a military context are to be found the towers at Cardiff (Nash-Williams 1969, 70–2). Outside the area but also in a military context are the multangular towers on the river frontage of the fortress at York (RCHM 1962, 16–19).

6 eg Chichester (Wilson 1952, 166; Holmes 1962, 84) and London (Merrifield 1965, 111). One should note the bastions of mixed semicircular and rectangular shape at Caistor-by-Norwich; similar mixtures occur in the Saxon Shore fort at Richborough (Johnson 1976, 48), while other south-eastern Shore forts have variants of semicircular towers (Johnson 1976, 115).

7 eg Great Casterton (Corder 1961, 20–5) and Aldborough (Myers *et al* 1959, 46; Charlesworth 1971, 162). The outer faces of the corner bastions at Aldborough are, however, constructed on a shallow curve. The rectangular towers at Brough-on-Humber (Wacher 1969, 43) are in a possibly quasi-military context. There is also a little evidence to show that the southern side at least of the lower defences at Lincoln was provided with rectangular towers (Colyer & Jones 1979, 86). Exceptions, however, occur at Caistor and Horncastle, where supposedly semicircular towers were used; but the surviving masonry is, in almost every case, so mutilated that a degree of uncertainty must remain. Moreover, there are strong reasons for believing that both sites were more of a military than a civilian nature.

8 For much of its distance it was 0·43–0·61 m in width and 1·5 m high where it emerged through the gate (Hull 1958, 37, 87). Equally large were the sewers at Lincoln and Aosta, which at both places must have issued through the walls, possibly at several points.

9 eg the Aqua Marcia (Ashby 1935, 89). The aqueduct was built into the structure of the Porta San Lorenzo by Aurelian.

10 Investigations are still being carried out by the Lincoln Archaeological Research Committee until recently under the direction of the late M K Wood (*Britannia*, 7 (1976), 325; see also Whitwell 1970, 33).

11 The Raw Dykes aqueduct must have reached the town in the vicinity of the south gate. A possible pipe channel has been noted within the town running apparently northwards from this area (Wacher 1978, 345).

12 RCHM 1970, 585. The supply most probably entered the town near the west gate.

13 A probable water conduit was noted passing through the north-western triumphal arch; it presumably carried water from an aqueduct entering the town near or at the Chester Gate (Frere 1962, 154).

14 Wacher 1961, 67; 1964, 16. Another possible example is known at Dorchester, Oxon (Hogg & Stevens, 1937, 46).

References

Ashby, T, 1935 *The aqueducts of ancient Rome*, ed I A Richmond

Birley, R, 1977 *Vindolanda: a frontier post on Hadrian's Wall*

Blanchet, A, 1908 *Recherches sur les aqueducts et cloaques de la Gaul romaine*

Bogaers, J E, 1979 King Cogidubnus in Chichester: another reading of *RIB* 91, *Britannia*, **10**, 243–54

Brown, P D C, & McWhirr, A D, 1966 Cirencester 1965, *Antiq J*, **46**, 244–54

Charlesworth, D, 1971 The defences of Isurium Brigantum, in *Soldier and civilian in Roman Yorkshire* (ed R M Butler), 155–64
Colyer, C, & Jones, M J, 1979 Excavations at Lincoln. Second interim report: excavations in the lower town 1972–8, *Antiq J*, **59**, 50–91
Corder, P, 1955 The reorganisation of the defences of Romano-British towns in the fourth century, *Archaeol J*, **112**, 20–42
———, 1961 *The Roman town and villa at Great Casterton, Rutland. Third report*
Cullen, P R, 1970 Cirencester: the restoration of the Roman town wall, 1967–68, *Britannia*, **1**, 227–39
de Montauzan, G, 1908 *Les aqueducts antiques de Lyon*
Frere, S S, 1962 Excavations at Verulamium, 1961. Seventh and final interim report, *Antiq J*, **42**, 148–59
———, 1964a Verulamium, three Roman cities, *Antiquity*, **38**, 103–12
———, 1964b Verulamium then and now, *Bull Inst Archaeol Univ London*, **4**, 61–82
Grenier, A, 1931 *Manuel d'archéologie Gallo-Romaine*, **1**, Paris
———, 1960 *ibid*, **4**, 1
Hanson, W S, 1978 The Roman military timber supply, *Britannia*, **9**, 293–305
Hobley, B, 1967 An experimental reconstruction of a Roman military turf rampart, *Proc VII Internat Congress Roman Frontier Stud*, 46–139, Tel Aviv
———, 1975 The Lunt Roman fort and training school for Roman cavalry, Baginton, Warwicks, *Trans Birmingham Warwickshire Archaeol Soc*, **88**, 19–23
Hogg, A H A, & Stevens, C E, 1937 The defences of Roman Dorchester, *Oxoniensia*, **2**, 41–73
Holmes, J, 1962 The defences of Roman Chichester, *Sussex Archaeol Collect*, **100**, 80–110
Hull, M R, 1958 *Roman Colchester*
Johnson, S, 1976 *The Roman forts of the Saxon Shore*
Merrifield, R, 1965 *The Roman city of London*
Myers, J N L, Steer, K A, & Chitty, A M H, 1959 The defences of Isurium Brigantum (Aldborough), *Yorkshire Archaeol J*, **40**, 1–77
Nash-Williams, V E, 1930 Further excavations at Caerwent, Monmouthshire, 1923–5, *Archaeologia*, **80**, 229–88
———, 1969 *The Roman frontier in Wales*, 2 edn, rev M G Jarrett
Richmond, I A, 1930 *The city walls of imperial Rome*
RCHM, 1962 *An inventory of the historical monuments in the city of York*, **1**, *Eburacum, Roman York*, Royal Commission on Historical Monuments
RCHM, 1970 *Dorset*, **2**, *South-east*, part 3, Royal Commission on Historical Monuments
Todd, M, 1981 *The Roman town at Ancaster, Lincolnshire: the excavations of 1955–71*
von Petrikovits, H, 1971 Fortifications in the north-western Roman Empire from the third to the fifth centuries AD, *J Roman Stud*, **61**, 178–218
Wacher, J S, 1961 Cirencester, 1960. First interim report, *Antiq J*, **46**, 63–71
———, 1964 Cirencester, 1963. Fourth interim report, *ibid*, **44**, 9–18
———, 1969 *Excavations at Brough-on-Humber 1958–61*
———, 1975 *The towns of Roman Britain*
———, 1978 *Roman Britain*
Webster, G, & Hollingsworth, D, 1959 The Wroxeter aqueduct, *Trans Shropshire Archaeol Natur Hist Soc*, **56**, 133–7
Wheeler, R E M, & Wheeler, T V, 1936 *Verulamium: a Belgic and two Roman cities*
Whitwell, J B, 1970 *Roman Lincolnshire*
Wilkes, J J, 1969 *Dalmatia*
Wilson, A E, 1952 Chichester excavations, 1947–50, *Sussex Archaeol Collect*, **90**, 164–220
Wilson, D R, 1973 Roman Britain in 1972. I. Sites explored, *Britannia*, **4**, 270–323

Index

compiled by
L Merrifield

References are given to page column (a for left-hand column, b for right-hand column) eg 25a, 25b. References to figures are in bold type eg **25,** and refer to the page on which the figure is to be found, not to the figure number. For places outside Britain, the following abbreviations have been used: Alg = Algeria, Aus = Austria, Bel = Belgium, Fr = France, Ger = Germany, Gr = Greece, Hun = Hungary, It = Italy, Jor = Jordan, Lib = Libya, Mor = Morocco, Neth = Netherlands, Rum = Rumania, Sp = Spain, Sw = Switzerland, Syr = Syria, Tun = Tunisia, Tur = Turkey, Yug = Yugoslavia